Data-Driven Fault Detection for Industrial Processes

Zhiwen Chen

Data-Driven Fault Detection for Industrial Processes

Canonical Correlation Analysis and Projection Based Methods

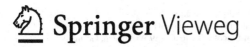 Springer Vieweg

Zhiwen Chen
Duisburg, Germany

Dissertation, University of Duisburg-Essen, 2016

ISBN 978-3-658-16755-4   ISBN 978-3-658-16756-1   (eBook)
DOI 10.1007/978-3-658-16756-1

Library of Congress Control Number: 2016961279

Springer Vieweg
© Springer Fachmedien Wiesbaden GmbH 2017

Printed on acid-free paper

This Springer Vieweg imprint is published by Springer Nature
The registered company is Springer Fachmedien Wiesbaden GmbH
The registered company address is: Abraham-Lincoln-Str. 46, 65189 Wiesbaden, Germany

To my parents, my love Caiwen and my son

Preface

In order to maximize the customer satisfaction and profit as well as to obey government regulations, the complexity and automation degree of modern industrial processes are significantly growing. To ensure the safety and overall reliability of such complicated processes, automatized fault detection is of great importance. Although the model-based fault detection theory has been well studied in the past decades, its applications are still limited for large-scale industrial processes because it is difficult to establish accurate model by means of first principles. On the other hand, sufficient (real-time) data are collected and recorded during process operations, and high-speed computation is available due to the rapid improvement in sensor and computer technologies. Therefore, the main objective of this work is to develop advanced data-driven fault detection methods for different application scopes.

This work is firstly dedicated to evaluate basic fault detection statistics with an alternative performance index. The commonly used T^2 and Q statistics are compared with respect to the geometric relationship and performance indices.

The further study focuses on developing effective fault detection methods for static and steady-state dynamic processes with available process input and output data. Different from the well-established methods based on multivariate analysis techniques, the core of the proposed methods is to build residual signals by means of the canonical correlation analysis technique for the fault detection purpose. However, the proposed methods are less powerful to deal with incipient multiplicative faults. To improve their fault detectability of such faults, the statistical local approach is integrated into the original methods.

For dynamic processes, an alternative fault detection method is proposed, in which residual signals are generated by means of a projection of process input data. This way of residual generation circumvents the parameter identification procedure. On the other hand, it also allows us to address deterministic disturbances, which are often not taken into account by the existing data-driven fault detection methods. Finally, the proposed fault detection methods are verified by four benchmark cases, i.e. the alumina evaporation process, the Tennessee Eastman process, the pilot scale continuous stirred tank heater and the inverted pendulum system. The application results shows the effectiveness of the proposed methods.

This work has been done at the Institute for Automatic Control and Complex Systems (AKS) in the Faculty of Engineering at the University of Duisburg-Essen, Germany. First of all, I owe the deepest gratitude to my supervisor, Prof. Dr.-Ing. Steven X. Ding, for all the inspiration and help he provided during the course of the last four years. I am heartily thankful for his guidance on my scientific research work. My sincere thanks must go to Prof. Dr. Ping Zhang for the valuable comments, which improved the quality of this work. I would like to express my sincere thanks to my group mates Dr.-Ing. Kai Zhang, Dr. Zhangming He and Dr.-Ing. Haiyang Hao for the valuable discussions and suggestions; Prof. Yuri Shardt for sharing his rich and valuable experience on academic research and scientific writing.

I would like to thank Dr.-Ing. Linlin Li, M.Sc. Changchen Xiang, Dr.-Ing. Dongmei Xu, Dr. Yong Zhang, M.Sc. Sihan Yu, Dr.-Ing. Hao Luo, M.Sc. Minjia Chang, Dr. Shasha Li, Dr.-Ing. Ying Wang for their help during my stay at AKS. My thanks should also go to all my other AKS colleagues, Dr.-Ing. Tim Könings, Dr. Zhiqiang Ge, Dr. Mingzhu Tang, Dr.-Ing. Chris Louen, Mrs. Sabine Bay, Dr.-Ing. Shane Dominic, Dr.-Ing. Köppen-Seliger, Dipl.-Ing. Eberhard Goldschmidt, Dr. Qingchao Jiang, M.Sc. Lu Qian, M.Sc. Changsheng Hua, M.Sc. Yunsong Xu, M.Sc. Zhengen Zhao, M.Sc. Tim Daszenies, M.Sc. Judith Minten, M.Sc. Svenja Siewers, Dipl.-Ing. Klaus Göbel and Mr. Ulrich Janzen, as well as the former colleagues M.Sc. Ping Liu, Prof. Yaguo Lei, Prof. Bo Shen, Prof. Hongli Dong, Prof. Xu Yang, Prof. Jianbin Qiu, Dr. Shouchao Zhai, Prof. Dr.-Ing. Shen Yin, Prof. Kaixiang Peng and Prof. Ying Yang for their valuable discussions and helpful suggestions. Without their help this work would not have been completed at this level.

I will forever be indebted to my big family, especially my parents and sisters, for all their support and love. In particularly, I would like to thank my wife, Caiwen Li, for all her support, encouragement, patience, and for being by my side.

My special thanks must go to Prof. Zhikun Hu, a wonderful mentor, who introduced me to 'fault diagnosis' and passed away during the writing of the work. He will be deeply missed.

Zhiwen Chen

Contents

List of Figures

List of Tables

List of Notations

Abbreviations

Abbreviation	Expansion
AEP	Alumina Evaporation Process
CCA	Canonical Correlation Analysis
CSTH	Continuous Stirred Tank Heater
CVA	Canonical Variate Analysis
DCCA	Dynamical Canonical Correlation Analysis
DCS	Distributed Control System
DO	Diagnostic Observer
DPCA	Dynamical Principal Component Analysis
DPLS	Dynamical Partial Least Squares
EVA	EVAporator
EVD	Eigen Value Decomposition
FAR	False Alarm Rate
FD	Fault Detection
FDF	Fault Detection Filter
FDR	Fault Detection Rate
GLR	Generalized Likelihood Ratio
I/O	Input/Output
IP	Inverted Pendulum
KR	Kernel Representation
LTI	Linear Time-Invariant
MC	Monte Carlo
MQC	Multivariate Quality Control
MVA	MultiVariate Analysis
MTFA	Mean Time to a False Alarm
NIPALS	Nonlinear Iterative PArtial Least Squares
PCA	Principal Component Analysis
PLS	Partial Least Squares
PMF	Probability Mass Function
PS	Parity Space
SCADA	Supervisory Control And Data Acquisition

SPE	Squared Prediction Error
SVD	Singular Value Decomposition
TE	Tennessee Eastman

Mathematical notations

Notation	Description
$\forall$	For all
$\in$	Belongs to
$\approx$	Approximately equal to
$:=$	Defined as
$>>$	Much greater than
$\Leftrightarrow$	Equivalent to
$\Rightarrow$	Implies
$\|\cdot\|_E$	Euclidean norm of a vector
x	A scalar
$\mathbf{x}$	A vector
$\hat{\mathbf{x}}$	Estimate of $\mathbf{x}$
$\mathbf{X}$	A matrix
$\mathbf{X}^T$	Transpose of $\mathbf{X}$
$\mathbf{X}^{-1}$	Inverse of $\mathbf{X}$
$\mathbf{X}^{\perp}$	Orthogonal complement of $\mathbf{X}$
$\mathbf{X}(:, p:q)$	a submatrix consisting of all the rows and the p-th to the q-th columns of $\mathbf{X}$
$\mathbf{I}_m$	m by m identity matrix
$\mathcal{R}$	The set of real numbers
$\mathcal{R}^n$	The set of n-dimensional real vectors
$\mathcal{R}^{n \times m}$	The set of $n \times m$ real matrices
$\text{eig}(\mathbf{X})$	Eigenvalue of $\mathbf{X}$
$\text{rank}(\mathbf{X})$	Rank of $\mathbf{X}$
$\text{tr}(\mathbf{X})$	Trace of $\mathbf{X}$
$\text{diag}(x_1, \ldots, x_n)$	A diagonal matrix formed with $x_1, \ldots, x_n$
$E(x)$ or $E(\mathbf{x})$	Expected value of x or $\mathbf{x}$
$\text{prob}(a > b)$	Probability that $a > b$
$\text{prob}(a > b|k)$	Probability that $a > b$ given k
$\mathcal{N}(\mu, \Sigma)$	Normal/Gaussian distribution with mean μ and covariance Σ
$\mathbf{x} \sim \mathcal{N}(\mu, \Sigma)$	$\mathbf{x}$ is distributed as $\mathcal{N}(\mu, \Sigma)$
J	Evaluation function

J_{th}	Threshold
$\mathcal{F}(k,l)$	$\mathcal{F}$-distribution with k and l degrees of freedom
$\chi^2(m)$	Chi-squared distribution with m degrees of freedom
α	Significance level
$\mathcal{F}_\alpha(k,l)$	Confidence limit satisfying $\mathrm{prob}(\mathcal{F}(k,l) > \mathcal{F}_\alpha(k,l)) = \alpha$
$\chi^2_\alpha(m)$	Confidence limit satisfying $\mathrm{prob}(\chi^2(m) > \chi^2_\alpha(m)) = \alpha$

1 Introduction

1.1 Background

In order to maximize customer satisfaction and profit as well as to satisfy government regulations, demands for process safety, product quality, economic operation and overall system reliability, are being pushed up in industry, e.g., in the process and manufacturing industry [74]. To meet these demands, the complexity and automation of modern industrial processes have significantly increased [26]. For example, in a typical alumina evaporation process, there are numerous control loops, in which a great number of sensors and actuators are embedded [14]. This development also brings great challenge in handling abnormal situations, such as performance degradation and component failure [106, 107, 108]. Based on data from insurance reports, it was estimated that the cost of lost production due to abnormal behavior is at least \$10 billion annually in the U.S. petrochemical industry.[1] In some extreme cases, the abnormality will further cause serious accidents. For example, in 2005, an explosion in a benzene production facility in Jilin Province, China, caused eight deaths and 60 injuries. The pollutant waste streams from the incident flowed into the Songhua River which is the main drinking water source, causing acute water pollution [13].

Fortunately, accidents are preventable if abnormal process conditions can be predicted and mitigated beforehand. Motivated by these observations, automatized fault detection (FD) is the essential step to ensure reliable treatment of undesirable events, which not only indicates the process upsets but also provides plant operators with practical assistance for reaction in time. Over the past decades, FD techniques have received considerable developments both in the academic and engineering domains, and are becoming an active research area in the control community, i.e., automatic control community and process control community.

1.2 Basic Concepts and Motivation of the Work

1.2.1 Basic Concepts of Fault Detection

It is quite common to ask what is a fault? Its definition varies across disciplines. In the control community, a *fault* in general is defined as an unpermitted deviation of at least

---

[1] See http://www.asmconsortium.net/defined/impact/Pages/default.aspx

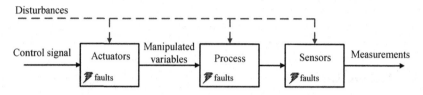

Figure 1.1: Schematic description of a standard industrial process

one characteristic property or parameter of the system from acceptable, usual, standard condition [40, 53]. In addition, it is well-accepted that the essential task of FD is [22]:

- the detection of the occurrence of faults in the functional units of the process, which lead to undesired or intolerable behavior (change) of the whole system.

Early detection may provide valuable warning on emerging abnormal behavior, so that appropriate measures can be taken to avoid (serious) accidents.

Figure 1.1 sketches a typical representation of industrial processes with automatic control systems. The sensors are the first element in the control loop to measure the process and output variables (measurements). These measurements provide essential process information to the controllers, in which the control signal is generated based on the operational setting. Usually, the control signal is converted for use by actuators. The outputs of actuators are defined as the manipulated variables, by which the process is operated, for instance, to change the input materials into the end products [99]. In reality, all these blocks are usually subject to stochastic disturbances (noises) and/or deterministic disturbances in the environment around the process. The primary task of a FD method is to detect faults under such noisy environment by using the measurements and/or the control signals, which contain all the information of faults. Note that using the manipulated variables instead of the control signals may be an advantage for FD methods when the actuators are highly nonlinear, because the required system equations do not contain the actuator nonlinearities [34].

1.2.2 Motivation for the Work

In this subsection, motivations of this work will be outlined. At first, the importance of data-driven FD methods will be described in comparison with other FD methods. It is followed by the description of problems to be solved in the data-driven framework, which also builds the study scope of this work. In what follows, the meaning of fault detection techniques and fault detection methods are the same.

Why Data-Driven Fault Detection Methods?

In engineering, the emergence of FD methods is strongly motivated by the concern of safety, reliability and cost of production. Their development is rapid and currently receiving considerable attention from both industry and academia. Numerous techniques have been proposed to solve fault detection problems in different industrial areas. According to the application scopes, existing FD techniques can be roughly classified into five categories:

- *Signal processing based techniques*: The signal processing based techniques have been extensively used in the mechanical engineering, where FD mainly severs for monitoring the operating condition [72, 79]. On the assumption that monitored process signals carry rich information about the faults, this information can be presented in the form of features or symptoms by signal processing. Typical features are time domain functions, including magnitudes, quadratic mean values, limit values, envelop, or frequency domain functions, such as spectral power densities and frequency spectral lines. The core idea of this type of technique is to extract fault information from process signals, based on it, a decision for a fault is made. Conventional techniques based on time- and frequency-domain analysis, e.g., the synchronous averaging [81] and Fourier analysis [82], are mainly used. Current research focuses on time-frequency analysis, e.g., wavelet analysis [60] and empirical mode decomposition [71]. However, except for the focus on one-block measurements, e.g., sensor measurements, the conventional signal processing based techniques are mainly valid for linear stationary processes [22].

- *Hardware redundancy based techniques*: Hardware redundancy is the duplication of critical components or functions of a system with the intention of increasing reliability of the system. This technique is widely used in the safety-critical areas like nuclear, aviation, and aerospace systems [41, 85]. The presence of faults is indicated if the output of the process component is different from the one of its identical (redundant) components. The fault can also be directly isolated under certain conditions. High reliability and direct fault isolation are the main appealing features of this technique. However, the application of this technique is also constrained by its high cost (redundant hardware) and the focus is on a small number of key components.

- *Qualitative model-based techniques*: The intuitive idea of the model-based FD techniques is to replace the hardware redundancy by a process model which is implemented in form of software. The qualitative model-based techniques, which are also known as knowledge-based methods, make use of qualitative models based on the available (*a priori*) process and fault knowledge [34, 106]. Different from the previously mentioned techniques, the basic knowledge for fault detection and diagnosis is usually a set of faults and the relationship between the features (symptoms) and the faults. The causal model-based techniques, including the signed digraphs and the fault tree analysis, have been well established [80, 86, 105]. The key idea behind these methods is to create the causal-effect relations between root nodes, which are presented in the form of variables in the signed digraphs and faults in the fault tree analysis. Application of the causal analysis can be difficult due to the increasing scale and complexity of the processes. The other practically effective techniques such as expert system and pattern recognition have also attracted a lot of attention. Although the qualitative model-based techniques are applicable to large-scale processes, the requirement of *a priori* knowledge of process and fault is a big constraint. Moreover, it is difficult to update or accommodate the case where new conditions are encountered due to lack of understanding of the physical process [117].

- *Quantitative model-based techniques*: Different from the qualitative model-based techniques, the redundancy, which is called analytical redundancy in the quantitative model-based techniques, is built based on the first principles process model. As shown in Figure 1.2, a basic model-based FD method consists of two steps, i.e., residual generation and residual evaluation. The residual signals are generated by comparing the outputs of the process and the quantitative model. In reality, processes are affected by disturbances and faults as described in Figure 1.1. Nevertheless, the quantitative model only represents the unaffected dynamic part. Under certain conditions, the residual signals contain all information about faults. Due to the influence of the disturbances and noise, further processing is needed. Usually, a residual evaluation function is designed to generate a scalar signal. Furthermore, a (constant) threshold should be determined based on the evaluation function and residual signals. Finally, a decision about the status of the process is made by comparing the scalar signal with the threshold. For example, if the evaluated residual is bigger than the threshold, it is said that the process has a fault; otherwise, the process is normal. The study of the quantitative model-based techniques was stimulated by Beard and Jones in the early 1970s. Nowadays, the well-known residual generation methods include the fault detection filter (FDF), diagnostic observer (DO), parity space (PS) approaches and the parameter identification-based approaches [43].

- Data-driven techniques: For many large-scale industrial processes, especially the chemical processes, building the previously mentioned quantitative model tends to be unfeasible or time-consuming and expensive in engineering. This limits the application of model-based FD methods. On the other hand, modern industrial processes are widely equipped with SCADA (supervisory control and data acquisition)

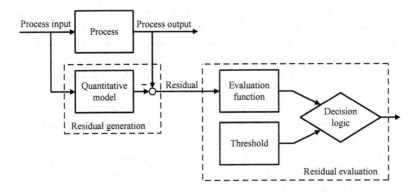

Figure 1.2: Basic quantitative model-based fault detection method

or DCS (distributed control system) systems, in which all possible signals are measured and stored [1]. This development has motivated researchers to extract models from the huge amount of historical process data and then to develop FD methods. Among existing methods, those based on multivariate analysis (MVA) technique have been established well. Using MVA technique for FD has been originally studied in the area of multivariate quality control (MQC) [54]. Typically, based on a statistical model built from the principal component analysis (PCA) technique, Hotelling's T^2 and the Q statistics are used to indicate faults in a complementary manner. The Q statistic is also known as the squared prediction error (SPE) statistic. However, the MQC methods mainly focus on the quality variables and the detection of quality relevant problems. In the early 1990s, the work of a research group [65, 76], led by MacGregor, first applied the partial least squares (PLS) technique to process variables, in addition to quality variables. This group is also well-known as the first ground-breaking work on multivariate statistical process monitoring (MSPM) field, in which MVA-based FD techniques play an essential role. Since then, the MVA-based approaches with PCA and PLS as its representative methods have been quickly developed and widely used due to their low engineering effort and high efficiency. Over the last two decades, various improvements to the conventional MVA-based methods have been made to deal with different types of processes, for instance dynamical [68], time-varying [32, 110] and nonlinear [20, 63]. In addition, some other methods have been introduced to MSPM field, such as independent component analysis (ICA)- [70, 95], and probabilistic MVA-based methods [120].

In practical applications, it may be difficult or expensive to derive the first principles model, especially for large-scale processes. To avert the difficulty in modeling and combine the power of quantitative model-based methods for dynamic processes, data-driven realization of the quantitative model-based techniques has become an active of research both in engineering and academia. It is well known that the well-established subspace identification method (SIM) can build the state space model from experimental measurements of the process [6]. Then, the conventional FD techniques are readily applicable [73, 100]. For FD purpose, however, the procedure of process model identification is often unnecessary [28]. Recently, with the aid of SIM [52, 93], the data-driven design (realization) of the quantitative model-based FD method has been achieved [26, 29]. The basic idea behind these methods is to directly construct the FD method using the collected process data without explicitly identifying system parameters. Furthermore, these methods are generalized in the kernel representation [24].

With the following discussions, the major issues and scopes addressed in this dissertation are outlined.

Why Performance Evaluation and Comparison of T^2 and Q Statistics?

From above analysis, it can be seen that data-driven FD methods have received considerable attention. Consequently, a great number of methods have been proposed in recent years [3, 67, 103, 116]. Therefore, an important issue is how to evaluate the performance of these FD methods for providing plant engineers and operators with guidance to select an appropriate FD method or develop new FD methods. For performance evaluation, the conventional way is to check false alarms and detection alarms. Two of commonly used evaluation indices are the false alarm rate (FAR) and the fault detection rate (FDR) [9, 92]. In addition, the time between the fault emergence and the first detection alarm, which called detection delay, is also an important performance of a FD method. The expected detection delay index, which is the expectation of detection delay, plays a complementary role to FDR [115]. In practice, the occurrence of false alarms is inevitable due to process uncertainties, such as the random disturbances, measurement error and inadequacy of model [12]. The FAR index only shows the overall performance of FD methods against these uncertainties, not how frequently false alarms arise. Note that frequent false alarms are neither convenient in practice because of the cost of stopping the production and searching for the root causes; nor desirable from psychological point of view, because the operator will stop using the FD system. Thus, as an alternative of FAR, the mean time to a false alarm (MTFA) is worth studying in order to meet practical demands.

In previous mentioned data-driven FD methods, T^2 and Q statistics are the most widely used ones. Note that the T^2 statistic is a generalized likelihood-ratio test statistic [23] and thus, following the Neyman-Pearson lemma, for a given significance level (acceptable FAR), the T^2 statistic gives the best fault detectability (FDR). The Q statistic is an alternative when the T^2 statistic is inaccurate due to numerical issues. Motivated by the widespread use of the two statistics, it is necessary to examine their FD performance. In addition, by comparison, some recommendations can be made for developing new FD methods, which would use the two statistics.

Why the Canonical Correlation Analysis-based Fault Detection Method?

Off-line training and on-line monitoring are two major steps in MVA-based FD methods, in which the PCA- and PLS-based ones are two representatives. However, the PCA-based methods only take into account the process variables in both steps. Although the PLS-based methods consider the process variables and output variables, in general, it is assumed that the latter variables are unmeasurable on-line or measurable with a large time delay. In this respect, the only available measurements in the on-line monitoring step are collected from process variables. As shown in Figure 1.1, the input-output relationship explicitly exists in standard industrial processes. In addition, with the improvement of sensor and routine data processing techniques, both variables are measurable on-line. Therefore, there is a need for making full use of the input-output relationship for fault detection.

As a classical technique in MVA, canonical correlation analysis (CCA) technique, which was originally proposed by Hotelling [49], is an efficient tool to find the correlation between two variables. From the viewpoint of fault detection, the correlation between input and output variables may provide a novel way in the statistical framework, because the dynamic (algebraic) relationship between them are well-developed in the quantitative model-based method. Motivated by these observations, it is worth developing the CCA-based FD methods as an extension of PCA-based or PLS-based methods for detecting faults in processes with input and output data.

Why the Detection of Incipient Multiplicative Faults?

In general, the variety of fault modes can be classified from different perspectives. As shown in Figure 1.1, faults are sorted into [22]:

- sensor faults: those faults that directly act on the process measurement

- actuator faults: those faults cause changes in the actuator

- process faults: those are used to indicate malfunctions within the process and can change the dynamical properties of the system

Based on the way the faults affect the process measurement, they can be classified as [45]:

- additive faults: those faults that only influence the mean value of the measurement for a stochastic system

- multiplicative faults: those faults that influence the variances, covariance, or higher-order statistical characteristics of the measurement for a stochastic system.

Based on the influence on system components [34], faults can be classified into as:

- abrupt faults: those faults occur like step changes

- incipient faults: those faults are usually slowly developing, e.g., drift and small changes in the variance

Roughly speaking, the three fault modes are not independent of each other. For instance, in Ding [22], offset and drift faults in sensors and actuators are called additive faults and malfunctions in the process or in the sensors and actuators which cause changes in the model parameters are called multiplicative faults.

Although a large amount of research work has focused on (abrupt) additive fault detection issues [47], studies on incipient multiplicative faults are relatively few [17]. The main difficulty in dealing with incipient faults is that their impact on the system components is slight. This slight impact such as inaccurate sensors or fouling on the heater transfer surface will gradually affect the process performance. Hence, it is desirable to develop a method for the early detection of this type of faults to improve process performance and prevent severe consequences.

Why Projection-based Fault Detection Method for Dynamic Processes with Deterministic Disturbances?

The MVA-based technique is generally applied to static or steady state dynamic processes, and shows optimal performance for fault detection in large-scale processes. For (general) dynamic processes with wide operating range, the data-driven kernel representation-based FD methods have been proposed [23, 24]. However, the existing methods, for example, the data-driven parity-space method, need a parameter identification procedure. For fault detection, a more direct way that using the orthogonal projection of process input data can avoid this identification procedure and be used to generate the residual signals. Moreover, in practice, the industrial processes are generally complex nonlinear systems that usually operate under certain operating conditions in the surrounding environment, in which the unknown deterministic disturbances commonly exist. The deterministic disturbances may cause high false alarm rate. The existing data-driven FD approaches are often not taken into account the deterministic disturbances. Fortunately, the alternative way for residual generation allows us to deal with this problem.

1.3 Objectives of the Dissertation

Based on above motivations, this dissertation first evaluates and compares the performance of two commonly used detection statistics, and then develops advanced FD methods from the perspective of process input and output variables, engineering effort and application scopes. To be specific, the objectives of this dissertation are:

- to define a new index for evaluating the performance of test statistics (in general, FD methods), and to study the relationship between the T^2 statistic and three cases of Q statistic and compare their FD performance with respect to the new and FDR indices;

- to propose CCA-based FD methods for static and steady state dynamic processes when the process input and output measurements are available;

- to improve CCA-based methods for detecting incipient multiplicative faults;

- to propose a projection-based FD approach for dynamic processes which can deal with deterministic disturbances;

Besides the theoretical contributions, industrial application is another objective of this dissertation. The effectiveness and applicability of the proposed methods are demonstrated on four benchmark cases.

1.4 Outline of the Dissertation

This dissertation is organized as shown in Figure 1.3. The major contributions of each chapter are summarized below.

Chapter 2: The Basics of Fault Detection

This chapter first introduces the technical descriptions of two kinds of industrial processes, i.e., static process and dynamic process, and models of different faults. Then, the basic FD statistics, i.e., T^2 and Q statistics, and FD methods such as PCA, PLS and DPCA are presented. In addition, the kernel representation-based FD systems, i.e., PS-based and DO-based ones, are reviewed for dynamic processes.

Chapter 3: Evaluation and Comparison of T^2 and Q Statistics for Fault Detection

In this chapter, the commonly used indices for evaluating the performance of FD methods are first reviewed. Since FAR only reflects the overall performance of a FD method against uncertainties, and not indicate how frequently false alarms will occur, the MTFA index is proposed to deal with this problem. Furthermore, the relationship between the commonly used statistics, i.e., T^2 and Q, is studied. It is followed by performance evaluation of these statistics in terms of the mean time to a false alarm and FDR. Finally, practical suggestions of using both statistics are given.

Chapter 4: Canonical Correlation Analysis-based Fault Detection Methods

In this chapter, CCA-based fault detection methods are proposed for both static and dynamic processes. Different from the well-established FD methods based on MVA techniques, the core of the proposed methods is to build residual signals by means of the CCA technique for the fault detection purpose. The effectiveness of the proposed methods will be demonstrated using numerical examples.

Chapter 5: Improved CCA-based Fault Detection Methods

In the last chapter, application of CCA to perform FD in both static and steady state dynamic processes is proposed. In dealing with incipient multiplicative faults, however, this method is less powerful. Thus, this chapter proposes improved CCA-based FD methods to detect incipient multiplicative faults by incorporating original CCA-based FD methods with the statistical local approach. The effectiveness of the method will be illustrated using a numerical example.

Chapter 6: A Projection-based Fault Detection Method for Dynamic Processes with Deterministic Disturbances

This chapter proposes an alternative data-driven FD method, in which the so-called residual signals are generated by means of a projection of process input data. This is the major difference to the existing model-based and data-driven FD approaches, where residual generator is realized based on the process input and output relationship/dynamics. Moreover, this way of residual generation avoids the parameter identification procedure and allows us to address deterministic disturbances, which are often not taken into account by data-driven FD methods.

Chapter 7: Benchmark Studies

In this chapter, methods presented in Chapter 4, 5 and 6 are demonstrated on four benchmark cases, the alumina evaporation process (AEP), the laboratory setup of continuous stirred tank heater (CSTH), the TE benchmark process and the inverted pendulum system. The effectiveness and applicability of these methods will be further checked.

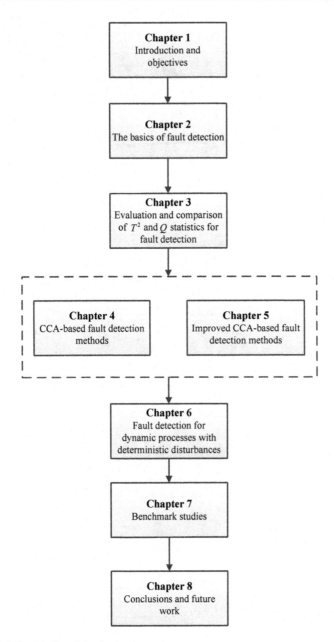

Figure 1.3: Organization of the chapters

2 The Basics of Fault Detection

As discussed in the introduction, the essential task of FD is to detect the presence of faults in the process. Such processes can be described by different types of system models, among which the linear time invariant (LTI) system is the mostly used one. According to the process dynamics and application scopes, such systems are commonly classified into static and dynamic classes. Therefore, this chapter is first dedicated to the mathematical descriptions of the two processes. Subsequently, the basic principle of fault detection is introduced. Moreover, an overview of the available data-driven FD methods, such as PCA- and PLS-based methods for static processes, and data-driven kernel representation-based methods for dynamic processes is given, which serves as the foundations of this dissertation.

2.1 Mathematical Descriptions of Industrial Processes

2.1.1 Representation of Static Processes

For those (steady) processes, the considered variables are expected to be stationary under normal operating condition. In this sense, the relationship between the input and output variables can be represented by a static process model. For example, in Hao *et al.* [45], the input is the low-level process variable and output is the high-level key performance indicators. A general representation of static processes can be described as [15]

$$\Psi_y \mathbf{y}(k) = \Psi_u \mathbf{u}(k) + \mathbf{v}(k), \qquad (2.1)$$

where $\Psi_y \in \mathcal{R}^{m \times m}$ and $\Psi_u \in \mathcal{R}^{m \times l}$ are constant, possibly unknown process parameter matrices, $\mathbf{u} \in \mathcal{R}^l$ is the input vector and $\mathbf{y} \in \mathcal{R}^m$ is the output vector, $\mathbf{v} \in \mathcal{R}^m$ is a normally distributed random vector with zero mean and (unknown) constant covariance Σ_v. In addition, $\mathbf{v}$ is uncorrelated with $\mathbf{u}$, i.e., $E(\mathbf{v}(\mathbf{u} - E(\mathbf{u}))^{\mathrm{T}}) = 0$. Conventionally, the process parameter matrix Ψ_y is considered to be the identity matrix in the application of PCA- and PLS-based methods [45, 121].

Assume the input and output vectors follow a multivariate normal distribution, denoted as

$$\mathbf{u} \sim \mathcal{N}(\mu_u, \Sigma_u), \ \mathbf{y} \sim \mathcal{N}(\mu_y, \Sigma_y),$$

where μ_u, Σ_u, μ_y and Σ_y are unknown but constant. As introduced in Section 1.2, the process of interest is usually subject to faults, which can be classified in different categories. We first consider the additive faults and multiplicative faults. In the data-driven framework, an additive fault which impacts the mean value of process variables, i.e., μ_u and μ_y, can be modeled by

$$\mathbf{u}_f(k) = \mathbf{u}(k) + \mathbf{f}_u(k), \quad \mathbf{y}_f(k) = \mathbf{y}(k) + \mathbf{f}_y(k)$$

where $\mathbf{f}_u(k)$ and $\mathbf{f}_y(k)$ represent those faults that affect the input variables and the output variables, respectively. If the considered fault is constant, the fault terms $\mathbf{f}_u(k)$ and $\mathbf{f}_y(k)$ are independent of the time instance k. Note that $\mathbf{f}_u$ and $\mathbf{f}_y$ can also be called actuator and sensor faults, respectively, according to the description in subsection 1.2.2. In practice, a multiplicative fault often cause changes in the model parameters such as $\boldsymbol{\Psi}_u$ and $\boldsymbol{\Psi}_y$, which could affect the covariance structure, i.e., Σ_u, Σ_y and Σ_ν. Generally, these multiplicative faults can be modeled as

$$\mathbf{y}_{mf}(k) = \mathbf{M}_y(k)(\mathbf{y}(k) - \mu_y) + \mu_y, \quad \mathbf{u}_{mf}(k) = \mathbf{M}_u(k)(\mathbf{u}(k) - \mu_u) + \mu_u$$

where $\mathbf{M}_u(k)$ and $\mathbf{M}_y(k)$ are matrices describing the influence of multiplicative faults. If the considered fault is constant, the two fault terms are constant matrices. For example, the element in the ith row and jth column, $M_y(i,j)$, representing the change of variance for $i = j$ and covariance for $i \neq j$, respectively.

In this dissertation, abrupt and incipient faults are also considered. Unless otherwise specified, all faults are assumed to be abrupt faults. As for incipient faults, especially incipient multiplicative faults, the change in model parameters is relatively slight.

2.1.2 Representation of Dynamic Processes

A dynamic process can be described by ordinary differential equations (ODEs). Considering that the subsequent study only focuses on developing the FD method in the data-driven fashion, the linear discrete time invariant model is used. A standard model form for such processes is the state space representation given by

$$\mathbf{x}(k+1) = \mathbf{A}\mathbf{x}(k) + \mathbf{B}\mathbf{u}(k), \quad \mathbf{x}(0) = \mathbf{x}_0, \tag{2.2}$$
$$\mathbf{y}(k) = \mathbf{C}\mathbf{x}(k) + \mathbf{D}\mathbf{u}(k) \tag{2.3}$$

where $\mathbf{x} \in \mathcal{R}^n$ is the state vector, $\mathbf{x}_0$ is the initial condition of the system, $\mathbf{u} \in \mathcal{R}^l$ and $\mathbf{y} \in \mathcal{R}^m$ are input and output vectors, respectively. Matrices $\mathbf{A}$, $\mathbf{B}$, $\mathbf{C}$ and $\mathbf{D}$ are real constant matrices with appropriate dimensions. If the process is corrupted by stochastic disturbances, the above state space representation is extended to

$$\mathbf{x}(k+1) = \mathbf{A}\mathbf{x}(k) + \mathbf{B}\mathbf{u}(k) + \boldsymbol{\eta}(k), \quad \mathbf{x}(0) = \mathbf{x}_0 \tag{2.4}$$
$$\mathbf{y}(k) = \mathbf{C}\mathbf{x}(k) + \mathbf{D}\mathbf{u}(k) + \boldsymbol{\varepsilon}(k), \tag{2.5}$$

where $\boldsymbol{\eta} \in \mathcal{R}^n$ and $\boldsymbol{\varepsilon} \in \mathcal{R}^m$ are assumed to be zero-mean, normally distributed white process noise (stochastic disturbances) satisfying

$$\mathbf{E}\left(\begin{bmatrix} \boldsymbol{\eta}(i) \\ \boldsymbol{\varepsilon}(i) \end{bmatrix} \begin{bmatrix} \boldsymbol{\eta}^{\mathrm{T}}(j) & \boldsymbol{\varepsilon}^{\mathrm{T}}(j) \end{bmatrix}\right) = \begin{bmatrix} \boldsymbol{\Sigma}_\eta & \boldsymbol{\Sigma}_{\eta\varepsilon} \\ \boldsymbol{\Sigma}_{\eta\varepsilon}^{\mathrm{T}} & \boldsymbol{\Sigma}_\varepsilon \end{bmatrix} \delta_{ij}, \quad \delta_{ij} = \begin{cases} 1, i = j \\ 0, i \neq j \end{cases} \tag{2.6}$$

which are statistically independent of the input vector $\mathbf{u}(k)$ and initial state vector $\mathbf{x}(0)$.

As discussed in subsection 1.2.2, the faults can be modeled in several ways. In case of the widely addressed additive faults, the system model (2.4)-(2.5) can be extended to

$$\mathbf{x}(k+1) = \mathbf{A}\mathbf{x}(k) + \mathbf{B}\mathbf{u}(k) + \mathbf{E}_f\mathbf{f}(k) + \boldsymbol{\eta}(k) \tag{2.7}$$

$$\mathbf{y}(k) = \mathbf{C}\mathbf{x}(k) + \mathbf{D}\mathbf{u}(k) + \mathbf{F}_f\mathbf{f}(k) + \boldsymbol{\varepsilon}(k) \tag{2.8}$$

where $\mathbf{f}(k) \in \mathcal{R}^{k_f}$ represents all types of possible faults. According to the faulty location, faults can be divided as sensor faults, actuator faults and process faults [22]. In this case,

- a sensor fault is modeled by setting $\mathbf{E}_f = \mathbf{0}$, $\mathbf{F}_f = \mathbf{I}$;

- an actuator fault is often formulated by setting $\mathbf{E}_f = \mathbf{B}$, $\mathbf{F}_f = \mathbf{D}$;

- a process fault can be modeled by $\mathbf{E}_f = \mathbf{E}_p$, $\mathbf{F}_f = \mathbf{F}_p$ for some $\mathbf{E}_p$ and $\mathbf{F}_p$.

It is worth noting that the additive faults only impact the mean value of $\mathbf{y}(k)$ and will not affect system stability.

Compared with additive faults, multiplicative faults are often modeled as changes of the system matrices, given by

$$\mathbf{x}(k+1) = (\mathbf{A} + \Delta\mathbf{A})\mathbf{x}(k) + (\mathbf{B} + \Delta\mathbf{B})\mathbf{u}(k) + \boldsymbol{\eta}(k), \tag{2.9}$$

$$\mathbf{y}(k) = (\mathbf{C} + \Delta\mathbf{C})\mathbf{x}(k) + (\mathbf{D} + \Delta\mathbf{D})\mathbf{u}(k) + \boldsymbol{\varepsilon}(k), \tag{2.10}$$

where $\Delta\mathbf{A}$, $\Delta\mathbf{B}$, $\Delta\mathbf{C}$ and $\Delta\mathbf{D}$ represent the influences of faults on the system matrices. These faults can cause changes in the second order statistics of the output data [44].

2.2 Basic Principle of Fault Detection

It has shown in subsection 1.2.2 that what all FD methods have in common is that a decision for a fault will be made based on the residual signals, which contains all fault information and uncertainties within or around the processes. In this subsection, we formally describe the basic FD principle through our work.

Denote the fault subspace by $\mathcal{D}_f$ and the (uncertain) environment subspace by $\mathcal{D}_\omega$. A fault $f \in \mathcal{D}_f$ can be a vector or a scalar variable, which can be time-varying or constant. $f = 0$ means fault-free case and $f \neq 0$ faulty cases. $\omega \in \mathcal{D}_\omega$ denotes noise or disturbance or some uncertainties in the process or the environment around the process. It can be either a vector or a matrix or a scalar variable.

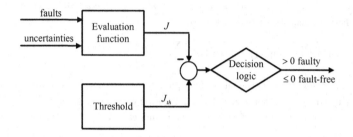

Figure 2.1: Schematic illustration of basic fault detection principle

Definition 2.1. *Let* $\mathcal{J} : \mathcal{D}_f \times \mathcal{D}_\omega \longrightarrow \mathcal{R}_+$ *denote an operator that maps a fault* $f \in \mathcal{D}_f$ *and* $\omega \in \mathcal{D}_\omega$ *to the evaluation function subspace* $\mathcal{R}_+$, *where symbol* $\times$ *denotes the Cartesian product and* $\mathcal{R}_+$ *represents the set with nonnegative (real) number. We call*

$$J = \mathcal{J}(f, \omega) \tag{2.11}$$

the evaluation function.

Remark 2.1. To simplify the decision rule, we assume in (2.11), without loss of generality, that $J \geq 0$. $\mathcal{J}$ is a general form of evaluation functions applied both in theoretical studies and applications and can be defined for static, dynamic, or other processes.

Essentially, a threshold, denote by J_{th}, is a bound of evaluation function J regarding to all possible process model uncertainties, disturbances and noises. On the assumption of a constant threshold $J_{th} > 0$, we consider in this dissertation the simplest form of a decision logic:

$$\begin{cases} J > J_{th} \Rightarrow \text{ faulty} \\ J \leq J_{th} \Rightarrow \text{ fault free.} \end{cases}$$

Figure 2.1 gives a schematic illustration of the basic fault detection principle of our work.

Without confusion, it should be noted that among a number of existing evaluation functions, in this dissertation, the one in statistical FD methods is called test statistic.

2.3 Basic Statistical Fault Detection Methods

In this section, basic statistical FD methods are reviewed. Conventionally, the process parameter matrix $\mathbf{\Psi}_y$ in the model (2.1) is considered as identity matrix. This leads to

$$\mathbf{y}(k) = \mathbf{\Psi}_u \mathbf{u}(k) + \mathbf{v}(k) \tag{2.12}$$

$$\mathbf{u} = \text{diag}(\sigma_{u,i}^{-1}, \ldots, \sigma_{u,l}^{-1})(\mathbf{u}_o - \text{E}(\mathbf{u}_o)) \tag{2.13}$$

$$\mathbf{y} = \text{diag}(\sigma_{y,j}^{-1}, \ldots, \sigma_{y,m}^{-1})(\mathbf{y}_0 - \text{E}(\mathbf{y}_o)) \tag{2.14}$$

where $\mathbf{y}$ and $\mathbf{u}$ are normalized in the form of (2.13) and (2.14), $\mathbf{u}_o$ and $\mathbf{y}_o$ denotes the observations, $\sigma_{u,i}$ and $\sigma_{y,j}$ represent the standard deviations of the ith input vector and the jth output vector. In order to make full use of the well-established statistical and probability theory, the normalization procedure plays a central role in constructing test statistics for statistical FD methods to make a decision for a fault. In the following subsections, the static model (2.12) instead of model (2.1) is used for reviewing the statistical FD methods. Additionally, in subsections 2.3.1 and 2.3.2, the measurements of $\mathbf{y}$ are only considered.

2.3.1 T^2 and Q Test Statistics

It is known that there are different types of test statistics applied for FD problems [23]. In this dissertation, we consider two commonly used types of test statistics: Hotelling's T^2-like and Q-like statistics. The introduction of them can be found as follows.

Hotelling's T^2 Statistic

For the sake of simplicity, let Hotelling's T^2 statistic be denoted by T^2. In our FD problems, two types of data (off-line data and on-line data) are used. As summarized in Anderson [5], T^2 is a statistic, in which the covariance of the data (measurement) is unknown and will be estimated by the data.

Suppose that the off-line data set $\mathbf{y}_{off}(k)$, $k = 1, \ldots, N_0$, and on-line data set $\mathbf{y}_{on}(k)$, $k = 1, \ldots, N_1$, follow a normal distribution with the same mean value and covariance matrix Σ_y. Let

$$\bar{\mathbf{y}}_0 = \frac{1}{N_0} \sum_{k=1}^{N_0} \mathbf{y}_{off}(k), \quad \bar{\mathbf{y}}_1 = \frac{1}{N_1} \sum_{k=1}^{N_1} \mathbf{y}_{on}(k)$$

be the mean value, and

$$\mathbf{S} = \frac{1}{N_0 + N_1 - 2} \left(\sum_{k=1}^{N_0} (\mathbf{y}_{off} - \bar{\mathbf{y}}_0)(\mathbf{y}_{off} - \bar{\mathbf{y}}_0)^{\text{T}} + \sum_{k=1}^{N_1} (\mathbf{y}_{on} - \bar{\mathbf{y}}_1)(\mathbf{y}_{on} - \bar{\mathbf{y}}_1)^{\text{T}} \right)$$

be the unbiased estimate of Σ_y, i.e., $\mathbf{S} \approx \Sigma_y$, then the T^2 test statistic for multiple samples is defined by

$$T_{ms}^2 = \frac{N_0 N_1}{N_0 + N_1} (\bar{\mathbf{y}}_1 - \bar{\mathbf{y}}_0)^{\text{T}} \mathbf{S}^{-1} (\bar{\mathbf{y}}_1 - \bar{\mathbf{y}}_0) \tag{2.15}$$

As analyzed in [78], T_{ms}^2 can be related to the $\mathcal{F}$-distribution by

$$T_{ms}^2 = (\bar{\mathbf{y}}_1 - \bar{\mathbf{y}}_0)^{\text{T}} \mathbf{S}^{-1} (\bar{\mathbf{y}}_1 - \bar{\mathbf{y}}_0) \sim \frac{m(N_0 + N_1 - 2)(N_0 + N_1)}{N_0 N_1 (N_0 + N_1 - m - 1)} \mathcal{F}(m, N_0 + N_1 - m - 1) \tag{2.16}$$

If only single on-line sample is used, i.e., $N_1 = 1$, then

$$\mathbf{S}_0 = \frac{1}{N_0 + 1 - 2} \left(\sum_{k=1}^{N_0} (\mathbf{y}_{off}(k) - \bar{\mathbf{y}}_0)(\mathbf{y}_{off}(k) - \bar{\mathbf{y}}_0)^{\mathrm{T}} + 0 \right) = \mathbf{S}$$

The test statistic for single on-line sample and the corresponding threshold are

$$T^2 = (\mathbf{y}_{on}(k) - \bar{\mathbf{y}}_0)^{\mathrm{T}} \mathbf{S}_0^{-1} (\mathbf{y}_{on}(k) - \bar{\mathbf{y}}_0) \tag{2.17}$$

$$J_{th,T^2} = \frac{m(N_0^2 - 1)}{N_0(N_0 - m)} \mathcal{F}_\alpha(m, N_0 - m) \tag{2.18}$$

In general, the number of the off-line data is sufficiently large, as analyzed in [104], the T^2 statistic approaches a χ^2-distribution. Hence, the corresponding threshold becomes

$$J_{th,T^2} = \chi_\alpha^2(m) \tag{2.19}$$

In what follows, the T^2 statistic for single on-line sample is used and the number of the off-line data is assumed to be sufficiently large. Thus, the threshold in (2.19) is adopted.

Q Statistic

The other common test statistic is the Q statistic, which is defined by

$$Q = (\mathbf{y}_{on}(k) - \bar{\mathbf{y}}_0)^{\mathrm{T}} (\mathbf{y}_{on}(k) - \bar{\mathbf{y}}_0) \tag{2.20}$$

Compared with the T^2 statistic, it is evident that the Q statistic can avoid the possible numerical problem brought by the inverse of $\mathbf{S}_0^{-1}$ [55]. However, the threshold setting of the T^2 statistic is simple, which can be approached by standard distribution table, for example the one in (2.19) can be determined by the standard χ^2 distribution table for the given significance level α and m degrees of freedom or the table of the quantile function for χ^2 distribution [39]. Unlike the case for the T^2 statistic, the distribution of the Q statistic is not standard and its threshold cannot be approached by a standard distribution table. In general, there are a number of approximations for the distribution of the Q statistic. In Section 3.2, three common approximations will be studied.

2.3.2 Principal Component Analysis-based Method

So far, we have introduced two common types of statistics that are extensively used for statistical FD methods. In the following subsections, we introduce several commonly used methods.

PCA is a basic technique of MVA and plays an important role both in the research and application domains [46, 58]. Although PCA was originally developed for other purposes, it is widely applied to FD due to its simplicity and effectiveness [18, 27, 35, 51, 61, 97,

113, 119]. It is regarded as an alternative to directly using the T^2 statistic on the available measurement of $\mathbf{y}$ [56]. In this subsection, we introduce the basic form of PCA applied to FD. Generally, such an application consists of two procedures, i.e., the off-line training procedure and the on-line implementation procedure.

Given N observations of the measurable variable $\mathbf{y}$, then the training data can be denoted as

$$\mathbf{Y} = [\mathbf{y}(1) \ \ldots \ \mathbf{y}(N)] \in \mathcal{R}^{m \times N} \tag{2.21}$$

where $\mathbf{y}(i) \in \mathcal{R}^m, i = 1, \ldots, N$, represents a sample vector. The data matrix $\mathbf{Y}$ is normalized to zero mean and optionally to unit variance.

Essentially, PCA aims to find these linear operators $\mathbf{p}$ for $i = 1, \ldots, \gamma$, that is

$$\mathbf{p}_i = \underset{\mathbf{p}_i^T \mathbf{p}_i = 1, \mathbf{p}_i^T \mathbf{p}_j = 0 (i \neq j)}{\arg\max} \mathbf{p}_i^{*T} \frac{\mathbf{Y}_i \mathbf{Y}_i^T}{N - 1} \mathbf{p}_i^* \tag{2.22}$$

Usually, a linear operator $\mathbf{p}_i$ is called as a loading vector. In this sense, PCA extracts those γ loading vectors which represent the most significant variability in data. In general, there are two ways to solve the optimization problem (2.22). The first way is called nonlinear iterative partial least squares (NIPALS), the core step is to deflate the data matrix as $\mathbf{Y}_{i+1} = (\mathbf{I}_m - \mathbf{p}_i \mathbf{p}_i^T) \mathbf{Y}_i, i = 1, \ldots, \gamma, \mathbf{Y}_1 = \mathbf{Y}$. For details, the reader is referred to [58]. In the second way, $\mathbf{p}$ can be obtained by performing an eigenvalue decomposition (EVD) or singular value decomposition (SVD) on the covariance matrix. Denote $\mathbf{\Sigma}_y = \frac{1}{N-1} \mathbf{Y} \mathbf{Y}^T$, then

$$\mathbf{\Sigma}_y = \mathbf{P} \mathbf{\Lambda} \mathbf{P}^T = [\mathbf{P}_{pc} \ \mathbf{P}_{res}] \begin{bmatrix} \mathbf{\Lambda}_{pc} & 0 \\ 0 & \mathbf{\Lambda}_{res} \end{bmatrix} \begin{bmatrix} \mathbf{P}_{pc}^T \\ \mathbf{P}_{res}^T \end{bmatrix} \tag{2.23}$$

where $\mathbf{P}_{pc} = [\mathbf{p}_1, \ldots, \mathbf{p}_\gamma] \in \mathcal{R}^{m \times \gamma}$ and $\mathbf{P}_{res} = [\mathbf{p}_{\gamma+1}, \ldots, \mathbf{p}_m] \in \mathcal{R}^{m \times (m-\gamma)}$ consist of the loading vectors, known as the principal components and residual components, respectively; $\mathbf{\Lambda}_{pc} = \mathrm{diag}(\lambda_1, \ldots, \lambda_\gamma)$ and $\mathbf{\Lambda}_{res} = \mathrm{diag}(\lambda_{\gamma+1}, \ldots, \lambda_m)$ contain the corresponding eigenvalues, satisfying $\gamma_1 \geq \ldots \geq \lambda_\gamma >> \lambda_{\gamma+1} \geq \ldots \lambda_m$.

For fault detection, T_{pca}^2 and SPE_{pca} (squared prediction error) test statistics are used

$$T_{pca}^2 = \mathbf{y}^T \mathbf{P}_{pc} \mathbf{\Lambda}_{pc}^{-1} \mathbf{P}_{pc}^T \mathbf{y} \tag{2.24}$$

$$SPE_{pca} = \mathbf{y}^T (\mathbf{I}_m - \mathbf{P}_{pc} \mathbf{P}_{pc}^T) \mathbf{y} \tag{2.25}$$

$$= \mathbf{y}^T \mathbf{P}_{res} \mathbf{P}_{res}^T \mathbf{y}$$

It is evident that SPE_{pca} is indeed the Q test statistic. Since, under the multivariate normal distribution assumption, $T_{pca}^2 \sim \chi^2(\gamma)$, the threshold is given as

$$J_{th, T_{pca}^2} = \chi_\alpha^2(\gamma) \tag{2.26}$$

As introduced in subsection 2.3.1, the threshold for the Q statistic is based on an approximation of the distribution. For instance,

$$J_{th,SPE_{pca},1} = g\chi_\alpha^2(h), \; g = \frac{\mathrm{tr}(\Lambda_{res}^2)}{\mathrm{tr}(\Lambda_{res})}, \; h = \frac{\mathrm{tr}^2(\Lambda_{res})}{\mathrm{tr}\Lambda_{res}^2} \tag{2.27}$$

$$J_{th,SEP_{pca},2} = \theta_1 \left(\frac{c_\alpha\sqrt{2\theta_2 h_0^2}}{\theta_1} + 1 + \frac{\theta_2 h_0(h_0 - 1)}{\theta_1^2} \right)^{1/h_0} \tag{2.28}$$

$$\theta_i = \sum_{j=\gamma+1}^{m} \lambda_j^i, \; i = 1,2,3, \; h_0 = 1 - \frac{2\theta_1\theta_3}{3\theta_2^2}$$

where c_α represents the normal deviation corresponding to the upper $1 - \alpha$ percentile [56]. The application procedures of PCA-based fault detection are summarized in Algorithm 2.1.

Algorithm 2.1. *PCA-based fault detection*

Off-line design based on the process data $Y \in \mathcal{R}^{m \times N}$

S1: Center the process data to obtain Y.
S2: Compute P_{pc}, P_{res}, Λ_{pc} *and* Λ_{res} *according to (2.23).*
S3: Set the thresholds for J_{th,T_{pca}^2} *and* $J_{th,SPE_{pca}}$ *according to (2.26) and (2.28), respectively.*
On-line implementation based on single sample y_k
S4: Build the statistics T_{pca}^2, *and* SPE_{pca} *according to (2.24) and (2.25), respectively.*
S5: Check the decision logic:

$$\begin{cases} T_{pca}^2 > J_{th,T_{pca}^2} \; or \; SPE_{pca} > J_{th,SPE_{pca}} \Rightarrow faulty \\ T_{pca}^2 \leq J_{th,T_{pca}^2} \; and \; SPE_{pca} \leq J_{th,SPE_{pca}} \Rightarrow fault\text{-}free. \end{cases}$$

Remark 2.2. It is worthwhile noting that if the principal component $\gamma < m$, then the matrix $P_{pc} \in \mathcal{R}^{m \times \gamma}$ is rank deficient, i.e., rank$(P_{pc}) = \gamma < m$. From the fault detection viewpoint, the matrix P_{pc} is not 'all pass' for faults, that is, there exists $f \neq 0$ such that $P_{pc}^T f = 0$. This situation is caused by the artificial design of rank deficient matrix P_{pc}. In this sense, the SPE_{pca} statistic is used as a complementary statistic for T_{pca}^2. If the principal component γ equals m, then the T_{pca}^2 statistic reduces to Hotelling's T^2 statistic, that is, when $\gamma = m$

$$T_{pca}^2 = y^T P_{pc} \Lambda_{pc}^{-1} P_{pc}^T y$$
$$= y^T P \Lambda^{-1} P^T y$$
$$= y^T \Sigma_y^{-1} y = T^2$$

this characteristic is also called affine equivalence [48] and the SPE_{pca} statistic will not be used.

2.3.3 Partial Least Squares-based Method

PLS regression is well developed in the field of MVA to construct the relationship between two sets of variables, especially, in case that there is a high degree of collinearity (redundancy) among the variables. It was proposed by H. Wold [114] for solving the collinearity problem as an alternative to the least squares regression. Since then, this method has been widely used. The early work on application of PLS to FD can be found in [65, 76]. In this application, one of these two data sets is chosen from the process variables. Another data set is collected from the key product quality variables or performance indicators. To effectively detect process upsets one must use all the information contained in both data sets. However, many important quality variables, such as thickness of the strip steel in a hot strip mill process and polymer molecular weight in a polymers production process, are still only measurable off-line and unmeasurable on-line or available on an hourly or daily basis [91, 94]. Therefore, the application of PLS to FD generally consists of two procedures, i.e., the off-line training procedure with two data sets and the on-line implementation with available process samples.

Given N observations of the variable $\mathbf{u}$ i.e., the process variable and $\mathbf{y}$ i.e., the quality variable, then the training data can be denoted as

$$\mathbf{U} = [\mathbf{u}(1) \ \dots \ \mathbf{u}(N)] \in \mathcal{R}^{l \times N} \text{ and } \mathbf{Y} = [\mathbf{y}(1) \ \dots \ \mathbf{y}(N)] \in \mathcal{R}^{m \times N}$$

Both data matrices are normalized to zero mean and optionally to unit variance.

The core idea of the off-line algorithm is to recursively solve the following optimization problem for $i = 1, \dots, \gamma$

$$\boldsymbol{\omega}_i^* = \arg\max_{\boldsymbol{\omega}_i^T \boldsymbol{\omega}_i = 1} \left\| \mathbf{U}_i \mathbf{Y}^T \boldsymbol{\omega}_i \right\|_E, \tag{2.29}$$

$$\mathbf{w}_i = \frac{\mathbf{U}_i \mathbf{Y}^T \boldsymbol{\omega}_i^*}{\left\| \mathbf{U}_i \mathbf{Y}^T \boldsymbol{\omega}_i^* \right\|_E}, \ \mathbf{t}_i = \mathbf{w}_i^T \mathbf{U}_i, \ \mathbf{p}_i = \frac{\mathbf{U}_i \mathbf{t}_i^T}{\mathbf{t}_i \mathbf{t}_i^T}, \tag{2.30}$$

$$\mathbf{r}_i = \begin{cases} \mathbf{w}_1, i = 1 \\ \prod_{j=1}^{i-1} \left(\mathbf{I}_l - \mathbf{w}_j \mathbf{p}_j^T \right) \mathbf{w}_i, i > 1 \end{cases}, \ \mathbf{q}_i = \frac{\mathbf{Y} \mathbf{t}_i^T}{\mathbf{t}_i \mathbf{t}_i^T}, \tag{2.31}$$

$$\mathbf{U}_{i+1} = \mathbf{U}_i - \mathbf{p}_i \mathbf{t}_i^T \tag{2.32}$$

where γ is the prescribed number of latent variables. Based on above solutions, the following matrices can be constructed:

$$\mathbf{P} = [\mathbf{p}_1, \dots, \mathbf{p}_\gamma], \ \mathbf{Q} = [\mathbf{q}_1, \dots, \mathbf{q}_\gamma]$$

$$\mathbf{R} = [\mathbf{r}_1, \dots, \mathbf{r}_\gamma], \ \mathbf{T} = [\mathbf{p}_1, \dots, \mathbf{p}_\gamma] = \mathbf{R}^T \mathbf{U}$$

Then, $\mathbf{u}$ and $\mathbf{y}$ are respectively decomposed into

$$\mathbf{u} = \mathbf{P} \mathbf{R}^T \mathbf{u} + (\mathbf{I}_l - \mathbf{P} \mathbf{R}^T) \mathbf{u}$$

$$\mathbf{y} = \mathbf{Q} \mathbf{R}^T \mathbf{y} + (\mathbf{I}_m - \mathbf{Q} \mathbf{R}^T) \mathbf{y}$$

For detection of quality-related and -unrelated process abnormalities (faults), the on-line measurable variable $\mathbf{u}$ is used to build the following two test statistics

$$T_{pls}^2 = \mathbf{u}^T \mathbf{R} \left(\frac{\mathbf{T}^T \mathbf{T}}{N-1} \right)^{-1} \mathbf{R}^T \mathbf{u}, \tag{2.33}$$

$$SPE_{pls} = \mathbf{u}^T \left(\mathbf{I}_l - \mathbf{PR}^T \right) \mathbf{u} \tag{2.34}$$

with the corresponding thresholds given as

$$J_{th,T_{pls}^2} = \chi_\alpha^2(\gamma) \tag{2.35}$$

$$J_{th,SPE_{pls}} = g\chi_\alpha^2(h), \quad g = \frac{\text{tr}(\Sigma_s^2)}{\text{tr}(\Sigma_s)}, \quad h = \frac{\text{tr}^2(\Sigma_s)}{\text{tr}\Sigma_s^2} \tag{2.36}$$

where Σ_s denotes the covariance of vector $\left(\mathbf{I}_l - \mathbf{PR}^T \right) \mathbf{u}$. The application procedures of PLS-based fault detection is summarized in Algorithm 2.2.

Algorithm 2.2. *PLS-based fault detection*

Off-line design based on the process data $\mathbf{U} \in \mathcal{R}^{l \times N}$ *and* $\mathbf{Y} \in \mathcal{R}^{m \times N}$
S1: Center the process data to obtain $\mathbf{U}$ and $\mathbf{Y}$.
S2: Compute $\mathbf{P}$, $\mathbf{R}$ and $\mathbf{T}$ according to (2.29)-(2.32).
S3: Set the thresholds for J_{th,T_{pls}^2} and $J_{th,SPE_{pls}}$ according to (2.35) and (2.36), respectively.
On-line implementation based on single sample u_k
S4: Build the statistics T_{pls}^2, and SPE_{pls} according to (2.33) and (2.34), respectively.
S5: Check the decision logic:

$$\begin{cases} T_{pls}^2 > J_{th,T_{pls}^2} \text{ and } SPE_{pls} \leq J_{th,SPE_{pls}} \Rightarrow \text{ fault influences quality variable} \\ T_{pls}^2 \leq J_{th,T_{pls}^2} \text{ and } SPE_{pls} > J_{th,SPE_{pls}} \Rightarrow \text{ fault does not influence quality variable} \\ T_{pls}^2 > J_{th,T_{pls}^2} \text{ and } SPE_{pls} > J_{th,SPE_{pls}} \Rightarrow \text{ both kinds of faults occur} \\ T_{pls}^2 \leq J_{th,T_{pls}^2} \text{ and } SPE_{pls} \leq J_{th,SPE_{pls}} \Rightarrow \text{ fault-free} \end{cases}$$

2.3.4 Dynamical PCA-based Method

The previously discussed PCA-based FD methods implicitly assume that the observations at current instant are statistically independent from observations at past time instances [68]. Consider a process with (obvious) dynamic behavior, which can be roughly represented in terms of the serial correlation between the current observation vector and the previous h observations, the dynamic PCA (DPCA), which takes into account the serial correlations, is a natural extension of the standard PCA. Suppose that the training observations are available in the time interval $[k - N, k]$. The training data matrix is formed in the following manner,

$$\mathbf{Y}_{k,h} = [\mathbf{y}_h(k - N + h) \ \cdots \ \mathbf{y}_h(k)] \in \mathcal{R}^{m(h+1) \times (N-h+1)} \tag{2.37}$$

where

$$\mathbf{y}_h(k) = \begin{bmatrix} \mathbf{y}(k) \\ \vdots \\ \mathbf{y}(k-h) \end{bmatrix} \in \mathcal{R}^{m(h+1)}$$

is the new observation vector obtained by augmenting the actual observation vector with the previous h observations. The remaining off-line training and on-line monitoring steps are then identical with the ones given in the standard PCA. Note that dynamic PLS (DPLS) deals with the serial correlation in the similar way as DPCA. Further information can be found in [64, 123].

2.4 Quantitative Model-based Residual Generation

During the past several decades, the quantitative model-based FD methods have been greatly developed for dynamic processes. The residual generation is a key step in those methods. Representative residual generation methods are FDF-, DO- and PS-based ones.

2.4.1 FDF-based Residual Generation

Consider the process described by model (2.4)-(2.5), in order to detect abnormal changes, a full order state observer, known as FDF, can be constructed for residual generation, that is

$$\hat{\mathbf{x}}(k+1) = \mathbf{A}\hat{\mathbf{x}}(k) + \mathbf{B}\mathbf{u}(k) + \mathbf{L}\left(\mathbf{y}(k) - \mathbf{C}\hat{\mathbf{x}}(k) - \mathbf{D}\mathbf{u}(k)\right), \qquad (2.38)$$

$$\mathbf{r}(k) = \mathbf{V}\left(\mathbf{y}(k) - \mathbf{C}\hat{\mathbf{x}}(k) - \mathbf{D}\mathbf{u}(k)\right) \qquad (2.39)$$

where $\mathbf{L}$ is the observer gain matrix and chosen such that $\mathbf{A} - \mathbf{L}\mathbf{C}$ is stable, i.e., its eigenvalues are located inside the unit circle. Denote the estimation error by $\mathbf{e}(k) = \mathbf{x}(k) - \hat{\mathbf{x}}(k)$, the dynamics of FDF is described by

$$\mathbf{e}(k+1) = (\mathbf{A} - \mathbf{L}\mathbf{C})\mathbf{e}(k) + \mathbf{\eta}(k) - \mathbf{L}\mathbf{\varepsilon}(k),$$

$$\mathbf{r}(k) = \mathbf{V}\mathbf{C}\mathbf{e}(k) + \mathbf{\varepsilon}(k)$$

where $\mathbf{V}$, called the post-filter, is a free design parameters aiming to achieve high sensitivity to faults and robustness against disturbances. Note that in the disturbance-free case, $\lim_{k\to\infty} \mathbf{e}(k) = 0 \implies \lim_{k\to\infty} \mathbf{r}(k) = 0$, which indicates that the process running in fault-free condition. If a fault occurs, $\lim_{k\to\infty} \mathbf{r}(k) \neq 0$ indicates the presence of fault. In practice, however, disturbances are inevitable. Then, $\lim_{k\to\infty} \mathbf{r}(k) \neq 0$ cannot be used to make any decision. Thus, the residual signal $\mathbf{r}(k)$ should be evaluated with a certain threshold, which can be set by the known disturbance information.

2.4.2 DO-based Residual Generation

Unlike FDF, DO is a Luenberger type (output) observer described by

$$\mathbf{z}(k+1) = \mathbf{Gz}(k) + \mathbf{Hu}(k) + \mathbf{Ly}(k), \tag{2.40}$$

$$\mathbf{r}(k) = \mathbf{Vy}(k) - \mathbf{Wz}(k) - \mathbf{Qu}(k) \tag{2.41}$$

where $\mathbf{z} = \mathbf{Tx} \in \mathcal{R}^s$, s denotes the observer order and can be different from the system order n. The parameter matrices $\mathbf{G}$, $\mathbf{H}$, $\mathbf{L}$, $\mathbf{V}$, $\mathbf{W}$ and $\mathbf{Q}$ as well as $\mathbf{T}$ have to satisfy the Luenberger conditions,

- $\mathbf{G}$ is stable

- $\mathbf{TA} - \mathbf{GT} = \mathbf{LC}, \ \mathbf{H} = \mathbf{TB} - \mathbf{LD}$

- $\mathbf{VC} - \mathbf{WT} = 0, \ \mathbf{Q} = \mathbf{VD}$

Denote $\mathbf{e}(k) = \mathbf{Tx}(k) - \mathbf{z}(k)$, the dynamics of DO is governed by

$$\mathbf{e}(k+1) = \mathbf{Ge}(k) + \mathbf{T}\boldsymbol{\eta}(k) - \mathbf{L}\boldsymbol{\varepsilon}(k),$$

$$\mathbf{r}(k) = \mathbf{VWe}(k) + \mathbf{V}\boldsymbol{\varepsilon}(k)$$

Similar to Section 2.4, the residual signal $\mathbf{r}(k)$ converges to zero under the disturbance-free condition. In practice, a certain threshold should be set based on the information about disturbance.

2.4.3 PS-based Residual Generation

Although a state space representation is used for the purpose of residual generation, the parity relation, instead of an observer, establishes the core of this approach. Consider the process model as described in (2.4-2.5) and, assume there is no redundancy in the output, i.e., rank$(\mathbf{C}) = m$. Given the PS order s, the process of interest can be extended as

$$\mathbf{y}_s(k) = \boldsymbol{\Gamma}_s\mathbf{x}(k-s) + \mathbf{H}_{u,s}\mathbf{u}_s(k) + \mathbf{H}_{\eta,s}\boldsymbol{\eta}_s(k) + \boldsymbol{\varepsilon}_s(k) \tag{2.42}$$

where $\mathbf{y}_s(k)$, $\mathbf{u}_s(k)$, $\boldsymbol{\eta}_s(k)$ and $\boldsymbol{\varepsilon}_s(k)$ are constructed as $\boldsymbol{\pi}_s(k)$ with the following data structure

$$\boldsymbol{\pi}_s(k) = \begin{bmatrix} \boldsymbol{\pi}(k-s) \\ \boldsymbol{\pi}(k-s+1) \\ \cdots \\ \boldsymbol{\pi}(k) \end{bmatrix}$$

and

$$\boldsymbol{\Gamma}_s = \begin{bmatrix} \mathbf{C} \\ \mathbf{CA} \\ \vdots \\ \mathbf{CA}^s \end{bmatrix} \in \mathcal{R}^{(s+1)m \times n}, \quad \mathbf{H}_{u,s} = \begin{bmatrix} \mathbf{D} & 0 & \cdots & 0 \\ \mathbf{CB} & \mathbf{D} & \ddots & \vdots \\ \vdots & \ddots & \ddots & 0 \\ \mathbf{CA}^{s-1}\mathbf{B} & \cdots & \mathbf{CB} & \mathbf{D} \end{bmatrix}$$

$$\mathbf{H}_{\eta,s} = \begin{bmatrix} \mathbf{0} & \mathbf{0} & \cdots & \mathbf{0} \\ \mathbf{C} & \mathbf{0} & \ddots & \vdots \\ \vdots & \ddots & \ddots & \mathbf{0} \\ \mathbf{CA}^{s-1} & \cdots & \mathbf{C} & \mathbf{0} \end{bmatrix}$$

Note that the only unknown variable is $\mathbf{x}(k - s)$. As known from the linear control theory, for $s \geq n$, the following rank condition holds:

$$\text{rank}(\mathbf{\Gamma}_s) \leq n < \text{the row number of matrix } \mathbf{\Gamma}_s = (s + 1)m$$

This ensures that there exists at least a row vector $\mathbf{v}_s(\neq 0) \in \mathcal{R}^{(s+1)m}$ such that

$$\mathbf{v}_s \mathbf{\Gamma}_s = 0 \tag{2.43}$$

Hence, the PS-based residual generator can be built as

$$\mathbf{r}(k) = \mathbf{v}_s(\mathbf{y}_s(k) - \mathbf{H}_{u,s}\mathbf{u}_s(k)) \tag{2.44}$$

Note that in the stochastic disturbance-free case, i.e., $\mathbf{H}_{\eta,s}\mathbf{\eta}_s(k) + \mathbf{\varepsilon}_s(k) = 0$, if $\mathbf{v}_s$ satisfies condition (2.43), then we have $\mathbf{r}(k) = 0$. When deterministic disturbances exist, the parity vector $\mathbf{v}_s$ is expected to be determined in such a way that the deterministic disturbances are completely decoupled. In practice, if a perfect decoupling of the disturbance is infeasible, the influence of the disturbances on $\mathbf{r}(k)$ should be minimized.

2.4.4 Kernel Representation

In fact, FDF-, DO- and PS-based residual generator as well as all other types of LTI residual generators can be constructed in a general form with the aid of kernel representation of system (2.2)-(2.3) [23], which is defined as a stable linear system, denoted as $\mathcal{K}$, driven by $\mathbf{u}(z)$, $\mathbf{y}(z)$ and satisfying

$$\forall \mathbf{u}(z), \ \mathbf{r}(z) = \mathcal{K} \begin{bmatrix} \mathbf{u}(z) \\ \mathbf{y}(z) \end{bmatrix} = 0 \tag{2.45}$$

where z denotes the z-transformation operator. There are different realizations of the kernel representation, as shown in Ding [23].

For instance, consider system (2.4)-(2.5), the PS-based residual generator in (2.44) can be rewritten as

$$\mathbf{r}(k) = [-\mathbf{v}_s\mathbf{H}_{u,s} \quad \mathbf{v}_s] \begin{bmatrix} \mathbf{u}_s(k) \\ \mathbf{y}_s(k) \end{bmatrix}$$

In this sense, $[-\mathbf{v}_s\mathbf{H}_{u,s} \quad \mathbf{v}_s]$ can be viewed as a realization of the kernel representation of system (2.4)-(2.5).

Note that the PS-based residual generator is a non-recursive realization of kernel representation, while the FDF- and DO-based residual generators are realized in the recursive form.

2.5 Data-Driven Kernel Representation-based FD Methods

An accurate mathematical process model plays an essential role in the above mentioned residual generators. However, in modern large-scale industrial processes, derivation of such models based on first principles or system identification methods can be costly and time-consuming. Motivated by these observations, direct development of a FD method from the measured input/output (I/O) data has received increasing attention [26, 28].

2.5.1 Data-Driven Realization of the Kernel Representation

The identification of the kernel representation is the essential step in data-driven design of residual generator. To this end, rewrite model (2.42) as

$$\begin{bmatrix} \mathbf{U}_{k,s} \\ \mathbf{Y}_{k,s} \end{bmatrix} = \mathbf{\Phi}_s \begin{bmatrix} \mathbf{U}_{k,s} \\ \mathbf{X}_{k-s} \end{bmatrix} + \begin{bmatrix} 0 \\ \mathbf{H}_{\eta,s}\mathbf{\eta}_{k,s} + \mathbf{\varepsilon}_{k,s} \end{bmatrix}, \ \mathbf{\Phi}_s = \begin{bmatrix} \mathbf{I} & 0 \\ \mathbf{H}_{u,s} & \mathbf{\Gamma}_s \end{bmatrix} \tag{2.46}$$

where $\mathbf{U}_{k,s} \in \mathcal{R}^{(s+1)l \times N}$, $\mathbf{Y}_{k,s} \in \mathcal{R}^{(s+1)m \times N}$, $\mathbf{\eta}_{k,s}$ and $\mathbf{\varepsilon}_{k,s}$ are formed as $\mathbf{\Omega}_{k,s}$ with the following data structure

$$\mathbf{\Omega}_{k,s} = [\mathbf{\pi}_s(k) \quad \ldots \quad \mathbf{\pi}_s(k+N-1)] \tag{2.47}$$

and

$$\mathbf{X}_{k-s} = [\mathbf{x}(k-s) \quad \ldots \quad \mathbf{x}(k-s+N-1)]$$

Since $\mathbf{\Phi}_s \in \mathcal{R}^{(s+1)(m+l) \times (n+(s+1)l)}$ and for $s \geq n$, $(s+1)(m+l) > n+(s+1)l$, there exists $\mathbf{\Phi}_s^{\perp}$ such that

$$\mathbf{\Phi}_s^{\perp}\mathbf{\Phi}_s = 0, \mathbf{\Phi}_s \in \mathcal{R}^{((s+1)m-n) \times (s+1)(m+l)} \tag{2.48}$$

Recall the kernel representation (2.45), we have

$$\mathbf{\Phi}_s^{\perp} \begin{bmatrix} \mathbf{U}_{k,s} \\ \mathbf{Y}_{k,s} \end{bmatrix} = \mathbf{\Phi}_s^{\perp} \begin{bmatrix} 0 \\ \mathbf{H}_{\eta,s}\mathbf{\eta}_{k,s} + \mathbf{\varepsilon}_{k,s} \end{bmatrix}$$

In other words, $\mathbf{\Phi}_s^{\perp}$ is a data-driven realization of the kernel representation and viewed as a residual generator.

The remaining issue is to identify $\mathbf{\Phi}_s^{\perp}$. Let

$$\mathbf{Z}_{k-s-1,s} = \begin{bmatrix} \mathbf{U}_{k-s-1,s} \\ \mathbf{Y}_{k-s-1,s} \end{bmatrix}$$

Note that $\mathbf{Z}_{k-s-1,s}$ is uncorrelated with $\mathbf{\eta}_{k,s}$ and $\mathbf{\varepsilon}_{k,s}$, i.e.,

$$\frac{1}{N-1}\mathbf{\eta}_{k,s}\mathbf{Z}_{k-s-1,s}^{\mathrm{T}} \approx 0, \ \frac{1}{N-1}\mathbf{\varepsilon}_{k,s}\mathbf{Z}_{k-s-1,s}^{\mathrm{T}} \approx 0$$

Hence, it holds

$$\frac{1}{N-1}\begin{bmatrix} \mathbf{U}_{k,s} \\ \mathbf{Y}_{k,s} \end{bmatrix} \mathbf{Z}_{k-s-1,s}^{\mathrm{T}} \approx \boldsymbol{\Phi}_s \begin{bmatrix} \mathbf{U}_{k,s} \\ \mathbf{X}_{k-s} \end{bmatrix} \mathbf{Z}_{k-s-1,s}^{\mathrm{T}}$$

Assume that

$$\begin{bmatrix} \mathbf{U}_{k,s} \\ \mathbf{X}_{k-s} \end{bmatrix} \mathbf{Z}_{k-s-1,s}^{\mathrm{T}} \text{ is of full row rank.}$$

The condition $\boldsymbol{\Phi}_s^{\perp} \boldsymbol{\Phi}_s = 0$ yields

$$\boldsymbol{\Phi}_s^{\perp} \begin{bmatrix} \mathbf{U}_{k,s} \\ \mathbf{Y}_{k,s} \end{bmatrix} \frac{\mathbf{Z}_{k-s-1,s}^{\mathrm{T}}}{N-1} = 0 \tag{2.49}$$

Doing an SVD

$$\begin{bmatrix} \mathbf{U}_{k,s} \\ \mathbf{Y}_{k,s} \end{bmatrix} \frac{\mathbf{Z}_{k-s-1,s}^{\mathrm{T}}}{N-1} = [\mathbf{U}_1\ \mathbf{U}_2] \begin{bmatrix} \boldsymbol{\Sigma}_1 & 0 \\ 0 & \boldsymbol{\Sigma}_2 \end{bmatrix} \begin{bmatrix} \mathbf{V}_1^{\mathrm{T}} \\ \mathbf{V}_2^{\mathrm{T}} \end{bmatrix} \tag{2.50}$$

leads to

$$\boldsymbol{\Sigma}_2 \approx 0, \quad \boldsymbol{\Phi}_s^{\perp} = \mathbf{U}_2^{\mathrm{T}} \in \mathcal{R}^{((s+1)m-n)\times(s+1)(l+m)} \tag{2.51}$$

2.5.2 Data-Driven PS-based Method

Denote $\boldsymbol{\Phi}_s^{\perp} = \begin{bmatrix} \boldsymbol{\Phi}_{s,u}^{\perp} & \boldsymbol{\Phi}_{s,y}^{\perp} \end{bmatrix}$, $\boldsymbol{\Phi}_{s,y}^{\perp} \in \mathcal{R}^{(s+1)m-n} \times (s+1)m$. It is evident that

$$\boldsymbol{\Phi}_{s,y}^{\perp} \boldsymbol{\Gamma}_s = 0, \quad \boldsymbol{\Phi}_{s,u}^{\perp} = -\boldsymbol{\Phi}_{s,y}^{\perp} \mathbf{H}_{u,s} \tag{2.52}$$

Thus, $\boldsymbol{\Phi}_{s,y}^{\perp} = \boldsymbol{\Gamma}_s^{\perp}$ is the parity space. Furthermore, a residual generator can be built as follows

$$\mathbf{r}_{ps}(k) = \boldsymbol{\Phi}_{s,y}^{\perp} \mathbf{y}_s(k) - \boldsymbol{\Phi}_{s,u}^{\perp} \mathbf{u}_s(k) \tag{2.53}$$

$$= \boldsymbol{\Phi}_{s,y}^{\perp} (\mathbf{H}_{\eta,s} \boldsymbol{\eta}_s(k) + \boldsymbol{\varepsilon}_s(k)) \sim \mathcal{N}(0, \boldsymbol{\Sigma}_r) \tag{2.54}$$

where $\boldsymbol{\Sigma}_r$ denotes the covariance matrix of the residual signal and can be estimated from the training data.

On the assumption that $\boldsymbol{\eta}$ and $\boldsymbol{\varepsilon}$ follow a normal distribution, the following T_{ps}^2 test statistic can be built for residual evaluation

$$T_{ps}^2 = \mathbf{r}_{ps}^{\mathrm{T}} \boldsymbol{\Sigma}_r^{-1} \mathbf{r}_{ps} \tag{2.55}$$

and the corresponding threshold is determined by the χ^2-distribution as

$$J_{th,T_{ps}^2} = \chi_\alpha^2((s+1)l - n) \tag{2.56}$$

However, with the measurement data in the time interval $[k-s,k]$ as its input, the residual signal $\mathbf{r}_{ps}(k)$ may contain redundant information, so that $\boldsymbol{\Sigma}_r$ may be rank deficient. To avoid this problem, we can apply the Q_{ps} statistic for residual generation

$$Q_{ps} = \mathbf{r}_{ps}^{\mathrm{T}} \mathbf{r}_{ps} \tag{2.57}$$

Its threshold is determined as

$$J_{th,Q_{ps}} = g\chi_\alpha^2(h) \tag{2.58}$$

Algorithm 2.3 summarizes the data-driven PS-based FD method based on the realization of the kernel representation.

Algorithm 2.3. *Data-driven PS-based FD method*

S1: Determine design parameters s and construct process I/O data matrices $\mathbf{Z}_{k-s-1,s}$, $\mathbf{U}_{k,s}$ and $\mathbf{Y}_{k,s}$.
S2: Perform an SVD according to (2.50).
S3: Extract the kernel representation matrices $\mathbf{\Phi}_{s,u}^\perp$ and $\mathbf{\Phi}_{s,y}^\perp$.
S4: Set the thresholds according to (2.56) and (2.58).
On-line implementation based on $\mathbf{u}_s(k)$ and $\mathbf{y}_s(k)$
S5: Build the test statistic T_{ps}^2 or Q_{ps} based on need.
S6: Check the decision logic:

$$\begin{cases} \text{test statistic} > \text{threshold} \Rightarrow \text{faulty} \\ \text{otherwise} \Rightarrow \text{fault-free.} \end{cases}$$

2.5.3 Data-Driven DO-based Method

On one hand, the one-to-one relationship between PS and DO has been extensively studied in [22]. On the other hand, the design effort for the on-line implementation of the DO-based method is lower than the PS-based one. Hence, based on each single PS-based residual generator, an equivalent DO can be constructed. Let

$$\boldsymbol{\alpha}_s = [\boldsymbol{\alpha}_{s,0}, \boldsymbol{\alpha}_{s,1}, \ldots, \boldsymbol{\alpha}_{s,s}], \ \boldsymbol{\beta}_s = [\boldsymbol{\beta}_{s,0}, \boldsymbol{\beta}_{s,1}, \ldots, \boldsymbol{\beta}_{s,s}] \tag{2.59}$$

where $\boldsymbol{\alpha}_s \in \mathcal{R}^{1\times(s+1)l}$ denotes a parity vector that can be selected as any row of $\mathbf{\Phi}_{s,y}^\perp$; $\boldsymbol{\beta}_s \in \mathcal{R}^{1\times(s+1)m}$ is obtained from $\mathbf{\Phi}_{s,u}^\perp$.

With $\boldsymbol{\alpha}_s$ and $\boldsymbol{\beta}_s$, we are able to construct an equivalent DO as

$$\mathbf{z}(k+1) = \mathbf{G}\mathbf{z}(k) + \mathbf{H}\mathbf{u}(k) + \mathbf{L}\mathbf{y}(k), \tag{2.60}$$

$$\mathbf{r}(k) = \mathbf{v}\mathbf{y}(k) - \mathbf{w}\mathbf{z}(k) - \mathbf{q}\mathbf{u}(k) \tag{2.61}$$

Recall $\mathbf{z} = \mathbf{T}\mathbf{x} \in \mathcal{R}^s$, parameters $\mathbf{G}, \mathbf{H}, \mathbf{L}, \mathbf{v}, \mathbf{w}$ and $\mathbf{q}$ should satisfy the Luenberger conditions,

- $\mathbf{G}$ is stable

- $\mathbf{TA} - \mathbf{GT} = \mathbf{LC}, \ \mathbf{H} = \mathbf{TB} - \mathbf{LD}$

- $vC - wT = 0, \quad q = vD$

Define $\mathbf{w} = [0 \ldots 0 \ 1]$, it has been demonstrated that all these matrices can be constructed using $\boldsymbol{\alpha}_s$ and $\boldsymbol{\beta}_s$ as follows [23]:

$$\mathbf{G} = \begin{bmatrix} 0 & 0 & \ldots & 0 \\ 1 & 0 & \ldots & 0 \\ \vdots & \ddots & \ddots & \vdots \\ 0 & \ldots & 1 & 0 \end{bmatrix} \in \mathcal{R}^{s \times s}, \quad \mathbf{L} = \begin{bmatrix} \boldsymbol{\alpha}_{s,0} \\ \boldsymbol{\alpha}_{s,1} \\ \vdots \\ \boldsymbol{\alpha}_{s,s-1} \end{bmatrix} \tag{2.62}$$

$$\mathbf{T} = \begin{bmatrix} \boldsymbol{\alpha}_{s,1} & \boldsymbol{\alpha}_{s,2} & \ldots & \boldsymbol{\alpha}_{s,s-1} & \boldsymbol{\alpha}_{s,s} \\ \boldsymbol{\alpha}_{s,2} & \ldots & \ldots & \boldsymbol{\alpha}_{s,s} & 0 \\ \vdots & \ldots & \ldots & \vdots & \vdots \\ \boldsymbol{\alpha}_{s,s} & 0 & \ldots & \ldots & 0 \end{bmatrix} \begin{bmatrix} \mathbf{C} \\ \mathbf{CA} \\ \vdots \\ \mathbf{CA}^{s-1} \end{bmatrix} \tag{2.63}$$

Based on $\mathbf{T}, \mathbf{L}, \mathbf{v}, \mathbf{H} = \mathbf{TB} - \mathbf{LD}, \mathbf{q} = \mathbf{vD}$ and the equality that $\boldsymbol{\Phi}_{s,u}^{\perp} = -\boldsymbol{\Phi}_{s,y}^{\perp} \mathbf{H}_{u,s}$, $\mathbf{H}$ and $\mathbf{q}$ can be constructed by

$$\mathbf{H} = \begin{bmatrix} \boldsymbol{\beta}_{s,0} \\ \vdots \\ \boldsymbol{\beta}_{s,s-1} \end{bmatrix}, \quad \mathbf{q} = \boldsymbol{\beta}_{s,s} \tag{2.64}$$

As a result, the data-driven DO-based FD method is summarized in Algorithm 2.4.

Algorithm 2.4. *Data-driven DO-based FD method*

S1: Construct α_s and β_s based on the achieved kernel matrices $\boldsymbol{\Phi}_{s,y}^{\perp}$ and $\boldsymbol{\Phi}_{s,u}^{\perp}$, respectively.
S2: With $\mathbf{w} = [0 \ldots 0 \ 1]$, construct parameters $\mathbf{G}, \mathbf{L}, \mathbf{v}, \mathbf{H}$ and $\mathbf{q}$ according to (2.63) and (2.64).
S3: Set the thresholds for $J_{th,T_{do}^2} = \chi_{\alpha}^2(1)$.
On-line implementation based on single sample $u(k)$ and $y(k)$
S4: Run the residual generator $(2.60) - (2.61)$ and evaluate the residual by the statistic

$$T_{do}^2 = r(k)^2 / \delta_r^2,$$

where δ_r^2 is the estimated variance of residual signal $r(k)$.
S5: Check the decision logic:

$$\begin{cases} T_{do}^2 > J_{th,T_{do}^2} \Rightarrow faulty \\ T_{do}^2 \leq J_{th,T_{do}^2} \Rightarrow fault\text{-}free. \end{cases}$$

2.6 Concluding Remarks

This chapter has introduced the basics for FD methods. The underlying mathematical descriptions of industrial processes has first presented based on the process dynamics and application scopes. Due to their simplicity and ability for easy application, MVA techniques are effective tools for FD of large-scale processes. Depending on the available statistical information, T^2 and Q statistics have been widely used for FD. For better understanding, it is worth studying the performance evaluation and comparison of both statistics. In addition, practical suggestions should be given to guide practitioner to the use of the Q statistic. This issue is further addressed in Chapter 3.

When input-output relationship explicitly exists and the two blocks of data are on-line measurable, CCA as an alternative solution to PCA and PLS technique should be explored for FD. The related issues are further addressed in Chapter 4 and Chapter 5.

The MVA techniques are generally applied to static processes. Although some dynamical modifications have been made based on them, their performance is excellent when dynamic processes reach steady state. Motivated by these limitations of MVA-techniques and the kernel representation-based methods, great efforts have been made on developing FD methods based on data-driven realization of kernel representation for general dynamic processes. However, recent studies demonstrate that an alternative method which avoids identifying the kernel matrices is applicable in this general dynamic case. The details can be found in Chapter 6.

3 Evaluation and Comparison of T^2 and Q Statistics for Fault Detection

In industrial applications, performance of an FD method is generally evaluated by FAR and FDR [12, 19, 22, 98]. Recently, Zhang *et al.* [122] proposed an evaluation index called expected detection delay, which is applicable for evaluating T^2 and Q statistics. Aimed at alarm management, Yang *et al.* [115] have investigated the analytical probability distribution of the time to a false alarm (TFA) using the classic general likelihood-ratio and cumulative sum-based FD methods. However, it does not seem applicable for these statistics applied in this dissertation, i.e., T^2 and Q statistics, owing to their high complexities. Therefore, in the first part of this chapter, a new index called mean time to a false alarm (MTFA) will be proposed for practical use.

As shown in the last chapter, the T^2 statistic follows a known distribution, for example, the F or χ^2 distribution, which allows an easy determination of threshold. Note that, to calculate the T^2 statistic, the inverse of the covariance matrix should be available. However, in some cases, the ill-conditional covariance matrix can cause numerical trouble to calculate the inverse. As an alternative, the Q statistic was proposed to avoid this problem and has been extensively used in the MVA-based FD methods. However, there is no work focusing on comparing the two statistics. Thus, in the second part, we would like to discuss the geometric relationship between them. Furthermore, in the third part, FDR and MTFA indices will be applied to evaluate the two statistics. The results are shown using numerical example studies.

3.1 Performance Evaluation

3.1.1 FAR and FDR

For performance evaluation of FD methods, the conventional way is to check the false alarm and detection alarm. Figure 3.1 illustrates both types of alarms. Given the time instance of the onset of a fault, a false alarm is understood as the event where an alarm is triggered when there is no fault, and a detection alarm is regarded as the event where an alarm arises when a fault actually occurs. From the probability point of view, FAR and FDR can be correspondingly defined as the probabilities of false alarms and detection alarms.

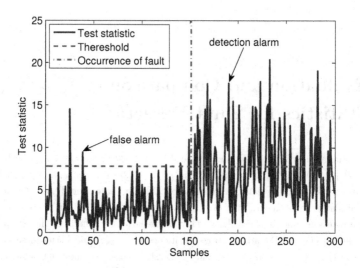

Figure 3.1: Illustration of false alarm and detection alarm

Recall that J and J_{th} denote the test statistic and corresponding threshold, the definitions of FAR and FDR can be shown as follows [22]

Definition 3.1. *Given J and J_{th}, we call the conditional probability*

$$FAR = prob(J > J_{th}|f = 0) \tag{3.1}$$

the false alarm rate.

Definition 3.2. *Given J and J_{th}, we call the conditional probability*

$$FDR = prob(J > J_{th}|f \neq 0) \tag{3.2}$$

the fault detection rate.

For calculation of FAR and FDR, there exist numerous approaches. For example, a numerical approximation approach was given in [118], which is simple for use but may has a large estimation error, and a theoretical approach was proposed in [122], which is based on a probabilistic framework but more involved for practical use. In this study, the numerical approximation approach based on randomized algorithms is used [102], which is simple for practical use and maintains an acceptable estimation error.

Although extensively used, FAR can only reflect the overall performance of test statistics against uncertainties. In practice, the process operator may be interested in how frequently false alarms occur. Therefore, in the next subsection, a new index MTFA in the probabilistic framework is proposed to investigate TFA.

3.1.2 Mean Time to A False Alarm

Let $\mathcal{T}$ denotes TFA, that is, the time to a false alarm. $\mathcal{T}$ may take the value j, where $j = 1, 2, \ldots, \infty$. Assume that FAR is constant at any time instance, so that the distribution of TFA is independent of whether a false alarm occurs or not before that. Obviously,

$$\text{prob}(\mathcal{T} = j) = \text{FAR}(1 - \text{FAR})^{j-1} \tag{3.3}$$

it is easy to show that

$$\sum_{j=1}^{\infty} \text{prob}(\mathcal{T} = j) = \sum_{j=1}^{\infty} \text{FAR}(1 - \text{FAR})^{j-1} = 1$$

which can determine the probability mass function of TFA in this circumstance.

Definition 3.3. *Given TFA, we call the expectation*

$$\text{MTFA} = E(\text{TFA})$$

$$= \sum_{j=1}^{\infty} j \text{prob}(\mathcal{T} = j)$$

$$= \sum_{j=1}^{\infty} j \text{FAR}(1 - \text{FAR})^{j-1}$$

the mean time to a false alarm.

Denoting the partial sum as

$$S_n = \sum_{j=1}^{n} j \cdot \text{FAR}(1 - \text{FAR})^{j-1} = \frac{1 - (1 - \text{FAR})^n}{\text{FAR}}$$

As a result, we have

$$\text{MTFA} = \lim_{n \to \infty} S_n = \frac{1}{\text{FAR}} \tag{3.4}$$

From (3.4), it can be seen that MTFA has an inverse relationship with FAR, that is to say, a small FAR leads to a large MTFA, and vice versa. In the ideal case, FAR is expected that equals to the given significance level α, based on which the expected MTFA is defined by

$$\text{MTFA}_e = \frac{1}{FAR} \approx \frac{1}{\alpha} \tag{3.5}$$

Conventionally, MTFA is addressed on the assumption of the existence of process and measurement noise whose statistical properties such as the mean and variance are well defined deterministic variables. In real applications, the situation may be different. If necessary, a total MTFA corresponding to all normal cases of interest can be calculated by

$$\text{MTFA} = \sum_{i=1}^{n} (\text{MTFA}^i \text{prob}(p^i)) \tag{3.6}$$

where p is the parameter vector like mean, (co)-variance etc. It is assumed that p is a random variable with a known distribution and finitely many values, in this case, is n. MTFA^i denotes the mean time to a false alarm given by the ith normal case.

3.2 Geometric Relationship between T^2 and Q Statistics

We assume that

- $\mathbf{y} \in \mathcal{R}^m \sim \mathcal{N}(0, \boldsymbol{\Sigma}_y)$

- Covariance matrix $\boldsymbol{\Sigma}_y$ is regular

Recall the T^2 statistic and three cases of Q statistic under consideration as

$$T^2 = \mathbf{y}^\mathsf{T} \boldsymbol{\Sigma}_y^{-1} \mathbf{y} \sim \chi^2(m) \tag{3.7}$$

$$Q = \mathbf{y}^\mathsf{T} \mathbf{y} \sim \begin{cases} \operatorname{tr}(\boldsymbol{\Sigma}_y)\chi^2(1) \\ \lambda_1 \chi^2(m) \\ g\chi^2(h) \end{cases} \tag{3.8}$$

where $\operatorname{tr}(\boldsymbol{\Sigma}_y) = \sum_{i=1}^m \lambda_i$, $g = \frac{\operatorname{tr}(\boldsymbol{\Sigma}_y \boldsymbol{\Sigma}_y)}{\operatorname{tr}(\boldsymbol{\Sigma}_y)} = \frac{\sum_{i=1}^m \lambda_i^2}{\sum_{i=1}^m \lambda_i}$ and $h = \frac{[\operatorname{tr}(\boldsymbol{\Sigma}_y)]^2}{\operatorname{tr}(\boldsymbol{\Sigma}_y \boldsymbol{\Sigma}_y)} = \frac{(\sum_{i=1}^m \lambda_i)^2}{\sum_{i=1}^m \lambda_i^2}$ [11]. λ_i denotes the ith eigenvalue of $\boldsymbol{\Sigma}_y$ ($i = 1, \ldots, m$) and $\lambda_1 \geq \lambda_2 \geq \ldots \geq \lambda_m$. Note that if $\boldsymbol{\Sigma}_y$ has the same eigenvalues, that is, $\lambda_1 = \lambda_m$, then $T^2 = \lambda_1^{-1} \mathbf{y}^\mathsf{T} \mathbf{y}$, which is indeed a kind of Q statistic. Therefore, in this study, we further assume that $\lambda_1 > \lambda_m$.

Given a significance level α, the corresponding thresholds are determined as

$$J_{th,T^2} = \chi_\alpha^2(m)$$

$$J_{th,Q} = \begin{cases} \operatorname{tr}(\boldsymbol{\Sigma}_y)\chi_\alpha^2(1) \\ \lambda_1 \chi_\alpha^2(m) \\ g\chi_\alpha^2(h) \end{cases}$$

where the three thresholds of Q are obtained based on modifications of the χ^2-distribution with different degrees of freedom.[1]

It is well known that the quadratic form of T^2 ($T^2 \leq J_{th,T^2}$) defines an ellipse for $m = 2$ or a hyperellipsoid in high dimensional space for $m \geq 3$. The lengths of semiaxes are given by the positive square roots of the reciprocals of eigenvalues of $(\boldsymbol{\Sigma}_y \boldsymbol{\Sigma}_y^\mathsf{T})^{-1}$, for example, the largest semiaxis equals to $\sqrt{\frac{1}{\max \operatorname{eig}((\boldsymbol{\Sigma}_y \boldsymbol{\Sigma}_y^\mathsf{T})^{-1})}}$. Since $\boldsymbol{\Sigma}_y$ is symmetric, the lengths of semiaxes equal to the eigenvalue of $\boldsymbol{\Sigma}_y$, with the largest semiaxis equals to λ_1 and the smallest semiaxis to λ_m [33]. Meanwhile, the form of Q ($Q \leq J_{th,Q}$) defines a circle for $m = 2$ or a hypersphere in high dimensional space for $m \geq 3$.

However, in the theory of statistics, the boundary of the quadratic form (3.7) is given with threshold J_{th,T^2}, which is determined by certain level of significance α. In this sense,

---

[1]The approximation of Q, which is not based on χ^2-distribution, for example the one in (2.28), is not considered.

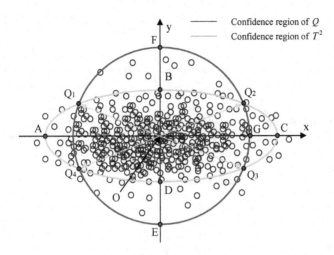

Figure 3.2: Confidence regions of T^2 and Q statistics

the largest semiaxis equals to $\lambda_1 J_{th,T^2}$ and the smallest semiaxis to $\lambda_m J_{th,T^2}$. For the quadratic form (3.8), the radius equals to the corresponding threshold $J_{th,Q}$.

For the sake of simplicity, we first consider the special case that $m = 2$. From Figure 3.2, by a given significance level α, it can be found out that

- the ellipse defines the confidence region of T^2, which is the set $\{T^2 | T^2 \leq J_{th,T^2}\}$. The x- and y-axes are represented by the eigenvectors of Σ_y.

- the circle defines the confidence region of Q, which is the set $\{Q | Q \leq J_{th,Q}\}$.

Apparently, the circle in red and ellipse in green intersect at four points, i.e., Q_1, Q_2, Q_3, Q_4, which leads to that the two confidence regions consist of an overlap region $(Q_1 Q_2 Q_3 Q_4)$ and four separated regions. The FD performance of T^2 and Q statistics with respect to MTFA and FDR, depend on the relationship between fault direction and the four regions, i.e., the upper region $(Q_1 F Q_2 B)$, the lower region $(Q_3 D Q_4 E)$, the left region $(Q_1 A Q_4)$ and the right region $(Q_2 C Q_3 G)$. For example, if a fault occurs along the upper and lower regions, the FDR of the T^2 statistic will larger than the one of the Q statistic, while the result is different when the fault directs to the left and right regions along the x-axis.

To link the preceding results to the comparison between the two statistics, one question is raised: do the circle and ellipse intersect with each other in four points for all three cases of the Q statistic?

To answer this question, we first give the formulae for the boundary curve of the corresponding confidence regions

$$\text{Ellipse: } \frac{y_1^2}{\lambda_1} + \frac{y_2^2}{\lambda_2} = J_{th,T^2} \tag{3.9}$$

$$\text{Circle: } y_1^2 + y_2^2 = J_{th,Q} \tag{3.10}$$

Let OA be the semimajor axis and OB the semiminor axis, then $OA = y_1$ and $OB = y_2$. It can be observed from Figure 3.2 and formulae (3.9-3.10) that the condition that the circle and ellipse have four intersections, i.e., A, B, C and D, should satisfies

$$OA^2 > J_{th,Q} \text{ and } OB^2 < J_{th,Q}$$

Note that

$$OA^2 = \lambda_1 J_{th,T^2}$$
$$OB^2 = \lambda_2 J_{th,T^2}$$

Therefore, the condition reduces to

$$\lambda_1 J_{th,T^2} > J_{th,Q} \tag{3.11}$$

$$\lambda_2 J_{th,T^2} < J_{th,Q} \tag{3.12}$$

It can be extended to the general case that $m > 2$. Then, conditions (3.11) and (3.12) are reformulated as

$$\lambda_1 J_{th,T2} > J_{th,Q} \tag{3.13}$$

$$\lambda_m J_{th,T2} < J_{th,Q} \tag{3.14}$$

which can guarantee that the m-dimensional spheroid and ellipsoid have intersections (not tangent).

In the following, we will discuss whether the three cases of the Q statistic satisfy conditions (3.13) and (3.14).

3.2.1 Distribution of Q is Approximated by $\lambda_1 \chi^2(m)$

In this case, $J_{th,T2} = \chi^2_\alpha(m)$, $J_{th,Q} = \lambda_1 \chi^2_\alpha(m)$, which gives

$$\lambda_1 J_{th,T2} = J_{th,Q}, \quad \lambda_m J_{th,T2} < J_{th,Q}$$

Thus, condition (3.13) fails. Indeed, the ellipse embeds in the circle. Therefore, the MTFA of the Q statistic will be higher than the one of the T^2 statistic, but for FDR, the situation is reversed.

3.2.2 Distribution of Q is Approximated by $\mathrm{tr}(\Sigma_y)\chi^2(1)$

We first consider a special case that $m = 2$. Since $J_{th,T2} = \chi_\alpha^2(2)$, $J_{th,Q} = (\lambda_1 + \lambda_2)\chi_\alpha^2(1)$, condition (3.11) turns into

$$\lambda_2 J_{th,T2} < J_{th,Q} \Rightarrow \lambda_2\chi_\alpha^2(2) < (\lambda_1 + \lambda_2)\chi_\alpha^2(1) \tag{3.15}$$

Due to $\lambda_2 < \lambda_1$ and $\chi_\alpha^2(2) < \chi_\alpha^2(1) + \chi_\alpha^2(1)$, we have

$$\lambda_2\chi_\alpha^2(2) < \lambda_2(\chi_\alpha^2(1) + \chi_\alpha^2(1)) < \lambda_1\chi_\alpha^2(1) + \lambda_2\chi_\alpha^2(1)$$

The condition (3.11) holds. Furthermore, condition (3.12) can be rewritten as

$$\lambda_1 J_{th,T2} > J_{th,Q} \Rightarrow \lambda_1\chi_\alpha^2(2) > (\lambda_1 + \lambda_2)\chi_\alpha^2(1) \tag{3.16}$$

It is difficult to determine whether condition (3.16) holds or not. Note that a counterexample is enough to deny this condition. Therefore, we take $\alpha = 0.05$ for example. Due to $\chi_\alpha^2(2)/\chi_\alpha^2(1) = 1.5597$ and $\chi_\alpha^2(2) < \chi_\alpha^2(1) + \chi_\alpha^2(1)$, condition (3.16) turns into

$$\lambda_1\frac{(\chi_\alpha^2(2)}{\chi_\alpha^2(1))} = 1.5597\lambda_1 > \lambda_1 + \lambda_2 \Rightarrow \frac{\lambda_1}{\lambda_2} > 1.7867$$

In this sense, the circle and ellipse do not overlap when $1 < \frac{\lambda_1}{\lambda_2} \leq 1.7867$. For other significance levels, e.g., 0.01, 0.02, 0.03 and 0.04, the conclusions are similar.

Therefore, in the general case that $m > 2$, it is harder to determine whether the hyperellipsoid and hypersphere intersects with four points, the results depend on the trace of the covariance matrix under consideration.

3.2.3 Distribution of Q is Approximated by $g\chi^2(h)$

First, reformulate condition (3.13-3.14) as

$$\lambda_1\chi_\alpha^2(m) > \frac{\lambda_1^2 + \lambda_2^2 + \ldots + \lambda_m^2}{\lambda_1 + \lambda_2 + \ldots + \lambda_m}\chi_\alpha^2(h) \tag{3.17}$$

$$\lambda_m\chi_\alpha^2(m) < \frac{\lambda_1^2 + \lambda_2^2 + \ldots + \lambda_m^2}{\lambda_1 + \lambda_2 + \ldots + \lambda_m}\chi_\alpha^2(h) \tag{3.18}$$

It can be shown that $1 < h < m$ due to the fact that

$$\frac{1}{m}(\lambda_1 + \lambda_2 + \ldots + \lambda_m)^2 < \lambda_1^2 + \lambda_2^2 + \ldots + \lambda_m^2 \tag{3.19}$$

Let $m = h + a$, $a \geq 0$, then $\chi_\alpha^2(m) \leq \chi_\alpha^2(h) + \chi_\alpha^2(a)$. Due to $\lambda_1(\lambda_1 + \lambda_2 + \ldots + \lambda_m) > (\lambda_1^2 + \lambda_2^2 + \ldots + \lambda_m^2)$, condition (3.17) is easy to be obtained that

$$\lambda_1\chi_\alpha^2(m) > \frac{\lambda_1^2+\lambda_2^2+\ldots+\lambda_m^2}{\lambda_1+\lambda_2+\ldots+\lambda_m}\chi_\alpha^2(h) \Rightarrow$$
$$\lambda_1(\lambda_1 + \lambda_2 + \ldots + \lambda_m)\chi_\alpha^2(m) > (\lambda_1^2 + \lambda_2^2 + \ldots + \lambda_m^2)\chi_\alpha^2(h)$$

From (3.18), we have

$$
\begin{aligned}
&\lambda_m \chi_\alpha^2(m) < \tfrac{\lambda_1^2+\lambda_2^2+\ldots+\lambda_m^2}{\lambda_1+\lambda_2+\ldots+\lambda_m}\chi_\alpha^2(h) \\
&\Rightarrow \tfrac{\lambda_1\lambda_m+\lambda_2\lambda_m+\ldots+\lambda_m^2}{\lambda_1^2+\lambda_2^2+\ldots+\lambda_m^2} < \tfrac{\chi_\alpha^2(h)}{\chi_\alpha^2(m)} \\
&\Leftrightarrow h\tfrac{\lambda_m}{\lambda_1+\lambda_2+\ldots+\lambda_m} < \tfrac{\chi_\alpha^2(h)}{\chi_\alpha^2(m)}
\end{aligned}
\tag{3.20}
$$

The remaining problem is to show inequality (3.20) holds.

Because there is no known analytic solution for $\chi_\alpha^2(m)$, it is hard to obtain a proof of (3.20). Thus, Algorithm 3.1 is designed to test this inequality. The achieved result based on this algorithm supports that inequality (3.20) holds.

Algorithm 3.1.

Step 1: Set $\alpha = [0.01, 0.02, 0.03, 0.04, 0.05, 0.06, 0.07, 0.08]$, m *to be any integer uniformly selected from the interval* $[1, 1000]$, n *is a randomly generated integer from* $[2, 100]$ *and the number of Monte Carlo (MC) experiments to be 50 million.*

Step 2: Generate m random eigenvalues selected uniformly from the interval $[\epsilon, 1000]$, *where ϵ is a very small real number.*

Step 3: Run the MC experiment according to (3.20) and record the results, in which sign 1 denotes inequality holds and sign 0 is the opposite.

Step 4: Check the decision logic:

$$
\begin{cases}
\text{no sign 0 exists} & \Rightarrow \text{ inequality (3.20) holds} \\
\text{otherwise} & \Rightarrow \text{ inequality (3.20) does not hold.}
\end{cases}
$$

From above discussions, only the third approximate distribution of the Q statistic guarantees the circle and ellipse intersects with four points. In this sense, when the information of certain type of fault is known, we can get the FDR of the Q statistic is higher than the one of the T^2 statistic. In practice, the information of fault is unknown and inverse of covariance matrix is feasible, so that the T^2 statistic is preferred under the demand of higher FDR. This point is demonstrated by a numerical example study in the next section.

3.3 Numerical Example Study

Two numerical example studies are provided in this section. The first one aims at showing the discussion results in the last section. The second one is used to compare the performance of the T^2 and Q statistics.

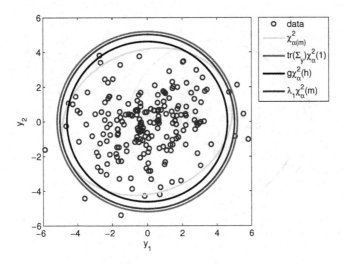

Figure 3.3: Confidence regions of T^2 and Q statistics in Example 1

3.3.1 Example for Geometric Relationship

In order to study the discussions in Section 3.2, the significance level is set to be 0.05. Two numerical examples are given as follows

- Example 1. Let $\mathbf{\Sigma}_y = \begin{bmatrix} 4 & 0.5 \\ 0.5 & 3 \end{bmatrix}$, then $\frac{\lambda_1}{\lambda_2} = 1.5064 < 1.7867$. Figure 3.3 shows that only the $g\chi_\alpha^2(h)$-based confidence region has four intersections with the $\chi_\alpha^2(m)$-based one.

- Example 2. Let $\mathbf{\Sigma}_y = \begin{bmatrix} 2 & 0.5 \\ 0.5 & 1 \end{bmatrix}$, then $\frac{\lambda_1}{\lambda_2} = 2.7836 > 1.7867$. The results can be seen in Figure 3.4 that both the $g\chi_\alpha^2(h)$- and $\mathrm{tr}(\mathbf{\Sigma}_y)\chi_\alpha^2(1)$-based confidence regions have four intersections with the one spanned by $\chi_\alpha^2(m)$.

The above two examples are consistent with the discussions in Section 3.2.

3.3.2 Example for Performance Evaluation

We now discuss the FD performance of the T^2 statistic and three cases of the Q statistic with respect to MTFA and FDR.

In the first part, the MTFA performance is compared. Let $m = 5$, then ten covariance matrices are generated by randomly setting,[2] and their eigenvalues are constrained from

---

[2]The Matlab code used to generate the covariance matrix is provided by Sihan Yu.

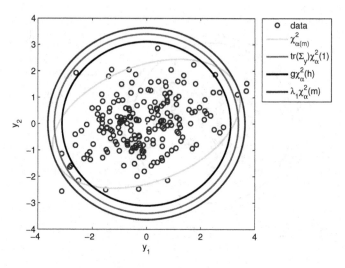

Figure 3.4: Confidence regions of T^2 and Q statistics in Example 2

0.1 to 100. Furthermore, 40,000 fault-free samples are generated based on these covariance matrices. For each covariance matrix, the MC experiment is run 20,000 times [102]. The MTFA is estimated using the mean value of the time to a false alarm. Assume that the probability of each covariance matrix is the same, as a result, the total MTFA is obtained according to (3.6). This comparison is conducted with eight level of significance, namely, 0.01, 0.02, 0.03, 0.04, 0.05, 0.06, 0.07, 0.08. Table 3.1 shows the comparison results, where Q_{max}, Q_{tr} and Q_{gh} represent the Q statistic, whose distribution is approximated by $\lambda_1\chi^2(m)$, $\mathrm{tr}(\Sigma_y)\chi^2(1)$ and $g\chi^2(h)$, respectively. In the second column, the expected MTFA is calculated according to (3.5). It can be seen from the fourth and fifth columns that the Q_{max} statistic has the largest MTFA, which is followed by the Q_{tr} statistic. Both of them are much larger than the expected MTFA. From the last two columns, we can see that the MTFAs of the Q_{gh} and T^2 statistics are around the expected MTFA. Furthermore, in the first four significance levels, the MTFA of the Q_{gh} statistic is less than the one of the T^2 statistic. In the last four significance levels, the MTFA of the Q_{gh} statistic is larger than the one of the T^2 statistic. From the eight cases, the MTFA values of the T^2 statistic are closer to the corresponding expected MTFA value due to its standard distribution. The one of the Q_{gh} statistic deviates a little from the expected MTFA value, because the distribution of it is not standard but approximated. In the second part, we study the detection performance. Assume that the fault structure is known and can be modeled as

$$\mathbf{y}_f = \mathbf{y} + \Xi f \qquad (3.21)$$

Table 3.1: MTFA for fault-free scenario

α	MTFA_e	Q_{max}	Q_{tr}	Q_{gh}	T^2
0.01	100.0	1420.32	909.14	82.53	100.19
0.02	50.00	1736.80	427.52	47.05	50.08
0.03	33.33	772.94	394.28	32.07	33.37
0.04	25.00	938.77	90.07	23.98	24.99
0.05	20.00	733.75	66.76	20.03	19.98
0.06	16.67	413.76	72.10	16.79	16.85
0.07	14.29	402.86	40.37	14.53	14.43
0.08	12.50	340.50	27.19	12.92	12.51

where Ξ represents the direction of fault and f denotes the magnitude of fault. In order to remove the influence of Ξ on the fault magnitude, the fault direction is scaled to unit length by $\frac{\Xi}{\|\Xi\|}$. Three fault magnitudes are considered, namely, 3, 5 and 8. In this part, two cases are considered as follows:

- In the first faulty case, the fault direction is assumed to be unknown, which is randomly generated. Eight directions are considered and with the same probability. The calculation of total FDR can be conducted in the similar way according to (3.6).

- In the second faulty case, the fault direction Ξ is assumed to be known *a priori*, which is set to be the first eigenvector of Σ_y

$$\Sigma_y = \begin{bmatrix} 4.86 & 0.41 & 0.51 & -1.05 & -0.57 \\ 0.41 & 5.98 & -1.56 & -1.73 & 2.09 \\ 0.51 & -1.56 & 2.79 & -1.10 & -1.62 \\ -1.05 & -1.73 & -1.10 & 2.60 & 0.70 \\ -0.57 & 2.09 & -1.62 & 0.70 & 5.10 \end{bmatrix}$$

For each fault magnitude, run the MC experiment 20,000 times. Table 3.2 shows the comparison results for FDR. In both cases, the Q_{max} statistic has the smallest FDR, which is followed by the Q_{tr} statistic. In the first case, the T^2 statistic has the largest FDR. In the second case, the Q_{gh} statistic has the largest FDR when the fault is along the direction of the first eigenvector, which represents the largest semiaxis.

Based on the performance of the four statistics in fault-free and fault scenarios, it is clear that the higher MTFAs of the Q_{max} and Q_{tr} statistics are at the cost of lower FDR. The T^2 statistic leads to a satisfactory tradeoff between the MTFA and the FDR. If the T^2 statistic is inaccurate due to numerical trouble, the Q_{gh} statistic is applicable as an alternative.

Table 3.2: FDR for fault scenario (%)

Ξ	f	Q_{max}	Q_{tr}	Q_{gh}	T^2
	3	1.55	3.08	13.29	$\boxed{25.03}$
unknown	5	7.02	11.73	32.96	$\boxed{56.65}$
	8	19.59	30.51	66.59	$\boxed{87.11}$
	3	2.34	4.16	$\boxed{14.69}$	10.04
known	5	9.05	13.89	$\boxed{33.71}$	21.70
	8	36.12	45.92	$\boxed{70.93}$	52.94

Remark 3.1. *Note that the T^2 statistic in this dissertation is constructed based on the unknown fault structure assumption. If the fault structure is known, the T^2 statistic can be reconstructed by taking into account this information [23].*

3.4 Concluding Remarks

This chapter has focused on evaluating and comparing the T^2 and Q statistics for fault detection. In the evaluation part, the conventional performance indices, FAR and FDR have been introduced. Motivated by the limit of FAR, an alternative index called MTFA has been propsoed, which can help process operators to learn how frequently false alarms will occur. It has been shown that the MTFA index has inverse relationship with FAR. By comparison studies, the geometric relationship between the T^2 and three cases of Q statistic, i.e., Q_{max}, Q_{tr} and Q_{gh} has been discussed. Only the confidence region given by the Q_{gh} statistic guarantees that it has four intersections with the one given by the T^2 statistic. This has led to that the performance of the Q_{max} and Q_{tr} statistics have undesired MTFA performance, and the FDR given by the Q_{gh} statistic may be higher than the one of the T^2 statistic in some special cases. This point has been verified by numerical example studies. The achieved results have revealed that the T^2 statistic has the desired MTFA performance and the highest FDR in the unknown fault structure case. The MTFA of the Q_{gh} statistic has relatively small deviations with the expected value compared with the Q_{max} and Q_{tr} statistics. In addition, at the significance level of 0.05, the Q_{gh} statistic has the smallest deviation. In the special faulty case, the Q_{gh} statistic has the highest FDR, even better than T^2. Finally, it can be concluded that *the T^2 statistic is the better choice for FD purpose, if it is unavailable, the alternative should be the Q_{gh} statistic.*

4 Canonical Correlation Analysis-based Fault Detection Methods

Statistical FD methods consist of two major procedures: off-line training and on-line monitoring. The differences between PCA- and PLS-based methods are that the PCA-based methods only consider the process variables in both procedures and detect changes in the condition of the process, sensors and actuators, while PLS-based methods are applied to process variables and output variables (quality variables) or key performance indicators, which are on-line unmeasurable or measurable only with a large time delay. In the off-line training procedure, quality data are used to guide the decomposition of the process data and to extract latent variables that are most relevant to the product quality. In the on-line monitoring procedure, only process variables are available and used to detect faults that are mostly related to the product quality variables or key performance indicators [122].

When input-output relationship explicitly exists and the two blocks of input and output data are on-line measurable, the CCA technique [49] is an efficient tool for developing a FD method. *CCA-based FD methods can be viewed as an extension of PCA-based or PLS-based methods for detecting faults in a process with input and output data.* Table 4.1 presents a comparison between CCA-, PCA-, and PLS-based methods to clarify the relationship between them. However, as a representative MVA technique, CCA has been rarely used for FD. Therefore, it is the purpose of this chapter to deal with FD issues using the CCA technique.

FD based on residual generation is the state of the art in the model-based FD framework [21]. Motivated by this, a canonical correlation-based residual generation is first realized by the CCA technique in the data-driven fashion. It is then applied to FD in static processes. In order to address FD in dynamic processes, the proposed static method is further extended to the dynamic version, which is similar to the well-established DPCA- and DPLS-based techniques.

4.1 Background and Problem Formulation

FD methods based on a quantitative model have been widely studied [21, 22, 38]. In these methods, residual generation is an essential step. As analyzed in Chapter 2, the kernel representation can parameterize all types of LTI residual generators when the process

Table 4.1: Tabular comparison between CCA-, PCA- and PLS-based FD methods

Method	Assumption on data	Variables	Detection purpose
PCA-based	Multivariate normal distribution	Process measurements (sensors)	Changes in sensor and process
PLS-based	Same as PCA, clear input-output relationship	Both input and output, and output is on-line unmeasurable	Changes related with quality variable
CCA-based	Same as PCA, clear input-output relationship	Both input and output, and both are on-line measurable	Changes in input and output as well as the process

input and output measurements are available. A time-domain realization of the kernel representation is applicable for residual generation, which is generally described in the following form

$$\mathbf{r}(k) = \mathcal{K}_u \mathbf{u}(k) + \mathcal{K}_y \mathbf{y}(k)$$

where $\mathbf{r}(k)$ is the residual signal at time k, $\mathbf{u}$ and $\mathbf{y}$ are input and output vectors, respectively. $[\mathcal{K}_u \ \mathcal{K}_y]$ represents the kernel representation of a nominal system under consideration. Note that $[\mathcal{K}_u \ \mathcal{K}_y]$ is usually known from the given process model. In this context, above kernel representation-based residual generation is conventionally referred as a model-based approach. Motivated by the facts that a process model is often not available or only achievable at high engineering cost, thus, the key step for generating the residual signals is to identify $\mathcal{K}_u$ and $\mathcal{K}_y$. Below, we address the residual generation problem in the data-driven fashion.

For static processes, the CCA technique is a useful tool to analyze the correlation between process input and output variables. Motivated by the kernel representation-based residual generation, in this chapter, the canonical correlation-based residual signal is defined by

$$\mathbf{r}_c(k) = \mathcal{L}^{\mathrm{T}} \mathbf{y}(k) - \mathcal{M}^{\mathrm{T}} \mathbf{u}(k)$$

where $\mathbf{r}_c(k)$ is the residual signal at time k. $\mathcal{L}$ and $\mathcal{M}$ are two constant matrices to be identified, respectively. An extension to dynamic processes by means of the DCCA-based method will be addressed in Section 4.4.

4.2 The Basic of CCA Technique

Suppose that the process under consideration has a input vector $\mathbf{u}_o \in \mathcal{R}^l$ and output vector $\mathbf{y}_o \in \mathcal{R}^m$. Assume that

$$\begin{bmatrix} \mathbf{u}_o \\ \mathbf{y}_o \end{bmatrix} \sim \mathcal{N}\left(\begin{bmatrix} \mu_u \\ \mu_y \end{bmatrix}, \begin{bmatrix} \boldsymbol{\Sigma}_u & \boldsymbol{\Sigma}_{uy} \\ \boldsymbol{\Sigma}_{uy}^{\mathrm{T}} & \boldsymbol{\Sigma}_y \end{bmatrix} \right) \tag{4.1}$$

where (cross-) covariance matrices $\boldsymbol{\Sigma}_u$, $\boldsymbol{\Sigma}_y$ and $\boldsymbol{\Sigma}_{uy}$ are regular. Denote the mean-centered input and output vectors, respectively, by $\mathbf{u}$ and $\mathbf{y}$, that is

$$\mathbf{u} = (\mathbf{u}_o - \mu_u) \sim \mathcal{N}(0, \boldsymbol{\Sigma}_u) \tag{4.2}$$

$$\mathbf{y} = (\mathbf{y}_o - \mu_y) \sim \mathcal{N}(0, \boldsymbol{\Sigma}_y) \tag{4.3}$$

Below, we introduce the standard CCA technique [5]. As the basis for the correlation evaluation, matrix

$$\boldsymbol{\Upsilon} = \boldsymbol{\Sigma}_u^{-1/2} \boldsymbol{\Sigma}_{uy} \boldsymbol{\Sigma}_y^{-1/2} \tag{4.4}$$

is first defined. Assume that

$$\mathrm{rank}(\boldsymbol{\Sigma}_{uy}) = \mathrm{rank}(\boldsymbol{\Upsilon}) = \kappa$$

Doing an SVD, the matrix $\boldsymbol{\Upsilon}$ can be decomposed as

$$\boldsymbol{\Upsilon} = \boldsymbol{\Gamma} \boldsymbol{\Sigma} \mathbf{R}^{\mathrm{T}} \tag{4.5}$$

with

$$\boldsymbol{\Gamma} = (\gamma_1, \dots, \gamma_l), \mathbf{R} = (\mathbf{r}_1, \dots, \mathbf{r}_m), \boldsymbol{\Sigma} = \begin{bmatrix} \boldsymbol{\Sigma}_\kappa & 0 \\ 0 & 0 \end{bmatrix}$$

where $\boldsymbol{\Sigma}_\kappa = diag(\rho_1, \dots, \rho_\kappa)$, $\rho_1 \geq \rho_2 \geq \dots \geq \rho_\kappa > 0$ are the singular values, which are also called canonical correlation coefficients [5]. γ_i, $i = 1, \dots, l$ and $\mathbf{r}_j$, $j = 1, \dots, m$ are the corresponding singular vectors.

Let

$$\mathbf{L} = \boldsymbol{\Sigma}_y^{-1/2} \mathbf{R} \in \mathcal{R}^{m \times m} \tag{4.6}$$

$$\mathbf{J} = \boldsymbol{\Sigma}_u^{-1/2} \boldsymbol{\Gamma} \in \mathcal{R}^{l \times l} \tag{4.7}$$

which are consist of the canonical correlation vectors [5]. It is evident that

$$\mathbf{L}^{\mathrm{T}} \boldsymbol{\Sigma}_y \mathbf{L} = \mathbf{I}_m, \quad \mathbf{J}^{\mathrm{T}} \boldsymbol{\Sigma}_u \mathbf{J} = \mathbf{I}_l \tag{4.8}$$

$$\mathbf{J}^{\mathrm{T}} \boldsymbol{\Sigma}_{uy} \mathbf{L} = \boldsymbol{\Sigma} \tag{4.9}$$

Based on the component κ, the parameters $\mathbf{J}$ and $\mathbf{L}$ can be decomposed into

$$\mathbf{L} = [\mathbf{L}_1, \mathbf{L}_2]$$

$$\mathbf{J} = [\mathbf{J}_1, \mathbf{J}_2]$$

where $\mathbf{L}_1 \in \mathcal{R}^{m \times \kappa}$, $\mathbf{L}_2 \in \mathcal{R}^{m \times (m-\kappa)}$, $\mathbf{J}_1 \in \mathcal{R}^{l \times \kappa}$ and $\mathbf{J}_2 \in \mathcal{R}^{l \times (l-\kappa)}$. Furthermore, we have

$$\mathbf{L}_1^T \boldsymbol{\Sigma}_y \mathbf{L}_1 = \mathbf{I}_\kappa, \quad \mathbf{J}_1^T \boldsymbol{\Sigma}_u \mathbf{J}_1 = \mathbf{I}_\kappa \tag{4.10}$$

$$\mathbf{J}_1^T \boldsymbol{\Sigma}_{uy} \mathbf{L}_1 = \boldsymbol{\Sigma}_\kappa, \quad \boldsymbol{\Sigma}_\kappa = \boldsymbol{\Sigma}_\kappa^T \tag{4.11}$$

We are now in a position to prove the following theorem, which shows important properties of CCA technique.

Theorem 4.1. *Given* $\boldsymbol{\Sigma}_u$, $\boldsymbol{\Sigma}_y$ *and* $\boldsymbol{\Sigma}_{uy}$, *if* $\mathbf{J}$, $\mathbf{L}$ *and* $\boldsymbol{\Sigma}$ *are obtained by doing an CCA on matrix* $\boldsymbol{\Sigma}_u^{-1/2} \boldsymbol{\Sigma}_{uy} \boldsymbol{\Sigma}_y^{-1/2}$, *then the following identities hold*

$$\mathbf{L}^T \boldsymbol{\Sigma}_{uy}^T = \boldsymbol{\Sigma}^T \mathbf{J}^T \boldsymbol{\Sigma}_u \tag{4.12}$$

$$\mathbf{J}^T \boldsymbol{\Sigma}_{uy} = \boldsymbol{\Sigma} \mathbf{L}^T \boldsymbol{\Sigma}_y \tag{4.13}$$

$$\mathbf{L}_1^T \boldsymbol{\Sigma}_{uy}^T = \boldsymbol{\Sigma}_\kappa \mathbf{J}_1^T \boldsymbol{\Sigma}_u \tag{4.14}$$

$$\mathbf{J}_1^T \boldsymbol{\Sigma}_{uy} = \boldsymbol{\Sigma}_\kappa \mathbf{L}_1^T \boldsymbol{\Sigma}_y \tag{4.15}$$

Proof. Assume that the identity (4.12) is not true, then

$$\mathbf{L}^T \boldsymbol{\Sigma}_{uy}^T \neq \boldsymbol{\Sigma}^T \mathbf{J}^T \boldsymbol{\Sigma}_u$$

Recall

$$\mathbf{J}^T \boldsymbol{\Sigma}_u \mathbf{J} = \mathbf{I}_l$$

So that multiplying the both sides of the inequality by $\mathbf{J}$ leads to

$$\mathbf{L}^T \boldsymbol{\Sigma}_{uy}^T \mathbf{J} = \boldsymbol{\Sigma}^T \neq \boldsymbol{\Sigma}^T \mathbf{J}^T \boldsymbol{\Sigma}_u \mathbf{J} = \boldsymbol{\Sigma}^T \Rightarrow \boldsymbol{\Sigma}^T \neq \boldsymbol{\Sigma}^T$$

It is in contradiction to such an identity. Thus, the assumption is wrong. That is, the identity (4.12) is true.

In a similar way, (4.13), (4.14) and (4.15) can also be proved. This completes the proof. □

4.3 CCA-based FD Method for Static Processes

4.3.1 CCA-based FD Method

This subsection is devoted to the development of a CCA-based FD method for static processes. Thus, the static model given in (2.1) is adopted.

Let $\mathbf{u}_{obs} \in \mathcal{R}^l$ and $\mathbf{y}_{obs} \in \mathcal{R}^m$ be the measured process input and output vectors, then they are centered with mean value and denoted as

$$\mathbf{u} = (\mathbf{u}_{obs} - \mu_u) \tag{4.16}$$

$$\mathbf{y} = (\mathbf{y}_{obs} - \mu_y) \tag{4.17}$$

Suppose that N samples of the process data are available and formed as

$$\mathbf{U} = [\mathbf{u}(1), \mathbf{u}(2), \ldots, \mathbf{u}(N)] \in \mathcal{R}^{l \times N}, \quad \mathbf{Y} = [\mathbf{y}(1), \mathbf{y}(2), \ldots, \mathbf{y}(N)] \in \mathcal{R}^{m \times N}$$

where $\mathbf{u}(i)$, and $\mathbf{y}(i)$, $i = 1, \ldots, N$ are the mean-centered input and output vectors as defined by (4.16) and (4.17) with the estimates

$$\mu_u \approx \frac{1}{N} \sum_{i=1}^{N} \mathbf{u}_{obs}(i), \quad \mu_y \approx \frac{1}{N} \sum_{i=1}^{N} \mathbf{y}_{obs}(i)$$

Furthermore, the covariances ($\boldsymbol{\Sigma}_u$, $\boldsymbol{\Sigma}_y$) and cross-covariance ($\boldsymbol{\Sigma}_{uy}$) of input and output can be estimated as

$$\boldsymbol{\Sigma}_u \approx \frac{1}{N-1} \sum_{i=1}^{N} (\mathbf{u}_{obs}(i) - \mu_u)(\mathbf{u}_{obs}(i) - \mu_u)^{\mathrm{T}} = \frac{\mathbf{U}\mathbf{U}^{\mathrm{T}}}{N-1}$$

$$\boldsymbol{\Sigma}_y \approx \frac{1}{N-1} \sum_{i=1}^{N} (\mathbf{y}_{obs}(i) - \mu_y)(\mathbf{y}_{obs}(i) - \mu_y)^{\mathrm{T}} = \frac{\mathbf{Y}\mathbf{Y}^{\mathrm{T}}}{N-1}$$

$$\boldsymbol{\Sigma}_{uy} \approx \frac{1}{N-1} \sum_{i=1}^{N} (\mathbf{u}_{obs}(i) - \mu_u)(\mathbf{y}_{obs}(i) - \mu_y)^{\mathrm{T}} = \frac{\mathbf{U}\mathbf{Y}^{\mathrm{T}}}{N-1}$$

By the CCA technique introduced in the last section, it turns out that

$$\left(\frac{\mathbf{U}\mathbf{U}^{\mathrm{T}}}{N-1}\right)^{-1/2} \left(\frac{\mathbf{U}\mathbf{Y}^{\mathrm{T}}}{N-1}\right) \left(\frac{\mathbf{Y}\mathbf{Y}^{\mathrm{T}}}{N-1}\right)^{-1/2} = \boldsymbol{\Sigma}_u^{-1/2} \boldsymbol{\Sigma}_{uy} \boldsymbol{\Sigma}_y^{-1/2}$$

$$= \boldsymbol{\Gamma}_s \boldsymbol{\Sigma} \boldsymbol{\Upsilon}_s^{\mathrm{T}}, \quad \boldsymbol{\Sigma} = \begin{bmatrix} \boldsymbol{\Sigma}_\varrho & 0 \\ 0 & 0 \end{bmatrix} \tag{4.18}$$

where $\text{rank}(\boldsymbol{\Sigma}_{uy}) = \varrho$, and $\boldsymbol{\Sigma}_\varrho = diag(\rho_1, \ldots, \rho_\varrho)$.

Let

$$\mathbf{J}_s = \boldsymbol{\Sigma}_u^{-1/2} \boldsymbol{\Gamma}(:, 1 : \varrho), \quad \mathbf{L}_s = \boldsymbol{\Sigma}_y^{-1/2} \boldsymbol{\Upsilon}(:, 1 : \varrho) \tag{4.19}$$

$$\mathbf{J}_{res} = \boldsymbol{\Sigma}_u^{-1/2} \boldsymbol{\Gamma}(:, \varrho + 1 : l), \quad \mathbf{L}_{res} = \boldsymbol{\Sigma}_y^{-1/2} \boldsymbol{\Upsilon}(:, \varrho + 1 : m) \tag{4.20}$$

Based on the identity (4.14) in Theorem 4.1 and the fact that noise exist in the process measurements, the correlation between $\mathbf{L}_s^{\mathrm{T}}\mathbf{y}$ and $\mathbf{J}_s^{\mathrm{T}}\mathbf{u}$ can be decomposed as

$$\mathbf{L}_s^{\mathrm{T}}\mathbf{y}(k) = \boldsymbol{\Sigma}_\varrho \mathbf{J}_s^{\mathrm{T}}\mathbf{u}(k) + \mathbf{v}_s(k)$$

$\mathbf{v}_s$ is the noise term, which is weakly correlated with $\mathbf{J}_s^{\mathrm{T}}\mathbf{u}$. As a result, we define residual vector as follows

$$\mathbf{r}(k) = \mathbf{L}_s^{\mathrm{T}}\mathbf{y}(k) - \mathbf{M}_s^{\mathrm{T}}\mathbf{u}(k) \tag{4.21}$$

where $\mathbf{M}_s^{\mathrm{T}} = \boldsymbol{\Sigma}_\varrho \mathbf{J}_s^{\mathrm{T}}$. It is known that the linear transformations of random vectors keep the same distribution, thus $\mathbf{r}$ also will follow a normal distribution.

Note that the covariance matrix of $\mathbf{r}(k)$ can be estimated by

$$\frac{1}{N-1}(\mathbf{L}_s^{\mathrm{T}}\mathbf{Y} - \boldsymbol{\Sigma}_\varrho \mathbf{J}_s^{\mathrm{T}}\mathbf{U})(\mathbf{L}_s^{\mathrm{T}}\mathbf{Y} - \boldsymbol{\Sigma}_\varrho \mathbf{J}_s^{\mathrm{T}}\mathbf{U})^{\mathrm{T}} = \frac{\mathbf{I}_\varrho - \boldsymbol{\Sigma}_\varrho^2}{N-1} \qquad (4.22)$$

We now address the issue of detecting faults in static processes. For such a detection purpose, it is reasonable to establish the following statistic

$$T_{cca}^2(k) = (N-1)\mathbf{r}^{\mathrm{T}}(k)(\mathbf{I}_\varrho - \boldsymbol{\Sigma}_\varrho^2)^{-1}\mathbf{r}(k) \qquad (4.23)$$

thus the threshold is determined as

$$J_{th,T_{cca}^2} = \chi_\alpha^2(\varrho) \qquad (4.24)$$

Note that if the two parts $\mathbf{L}_s^{\mathrm{T}}\mathbf{y}$ and $\mathbf{J}_s^{\mathrm{T}}\mathbf{u}$ are strongly correlated, that is, $\boldsymbol{\Sigma}_\varrho$ approaches the identity matrix $\mathbf{I}_\varrho$, then the $T_{cca}^2(k)$ value may be inaccurate due to the ill-conditioning matrix $(\mathbf{I}_\varrho - \boldsymbol{\Sigma}_\varrho^2)^{-1}$. In order to avoid this problem, as analyzed in Section 3.2, the Q statistic can be used as an alternative, which is

$$Q_{cca}(k) = \mathbf{r}^{\mathrm{T}}(k)\mathbf{r}(k) \qquad (4.25)$$

See discussions in Chapter 2, the threshold is set to be

$$J_{th,Q_{cca}} = g\chi_\alpha^2(h) \qquad (4.26)$$

where $g = S/2\mu$, $h = 2\mu^2/S$, S and μ are estimated as,

$$\mu = \frac{1}{N}\sum_{k=1}^{N} Q_{cca}(k), \quad S = \frac{1}{N-1}\sum_{k=1}^{N}(Q_{cca}(k) - \mu)^2$$

Note that the statistic T_{cca}^2 or Q_{cca} can detect those faults in measurement subspaces (u and y), which are correlated. In other cases, e.g. $\varrho < l$ or $\varrho < m$, the following two statistics can be built

$$T_u^2 = \mathbf{u}^T \mathbf{J}_{res}\mathbf{J}_{res}^T\mathbf{u} \qquad (4.27)$$

for detecting the faults in the u subspace, which is uncorrelated with y,

$$T_y^2 = \mathbf{y}^{\mathrm{T}}\mathbf{L}_{res}\mathbf{L}_{res}^{\mathrm{T}}\mathbf{y} \qquad (4.28)$$

for detecting the faults in the y subspace, which is uncorrelated with u.

We have $\mathbf{J}_{res}^T\mathbf{u} \sim \mathcal{N}(0, \mathbf{I}_{l-\varrho})$ and $\mathbf{L}_{res}^{\mathrm{T}}\mathbf{y} \sim \mathcal{N}(0, \mathbf{I}_{m-\varrho})$, so that corresponding thresholds are determined as

$$J_{th,T_u^2} = \chi_\alpha^2(l - \varrho)$$
$$J_{th,T_y^2} = \chi_\alpha^2(m - \varrho)$$

In Algorithm 4.1, the procedure of the CCA-based FD method for static processes is summarized. For the sake of simplicity, the Q_{cca} statistic is not included. If necessary, the T_{cca}^2 statistic can be replaced by the Q_{cca} statistic.

Algorithm 4.1. *CCA-based FD method*

Off-line design
S1: Center the process data to obtain $\mathbf{U}$ and $\mathbf{Y}$
S2: Perform an SVD on matrix $\Upsilon \Rightarrow \mathbf{J}_s, \mathbf{L}_s; \mathbf{J}_{res}; \mathbf{L}_{res}; \Sigma_\varrho$
S3: Compute the thresholds $J_{th,T^2_{cca}}$ and J_{th,T^2_y}
Online-implementation
S4: Build the statistics

$$
\begin{cases}
T^2_{cca} & \text{if } \varrho = l = m, \\
T^2_{cca}, T^2_u \text{ and } T^2_y & \text{other cases}
\end{cases}
$$

S5: Check the decision logic:

$$
\begin{cases}
T^2_{cca} > J_{th,T^2_{cca}} & \Rightarrow \text{ fault in } u \text{ and } y \text{ subspaces} \\
T^2_{cca} \leq J_{th,T^2_{cca}} \text{ and } T^2_u > J_{th,T^2_u} \Rightarrow \text{ fault in } u \text{ subspace, which is uncorrelated with } y \\
T^2_{cca} \leq J_{th,T^2_{cca}} \text{ and } T^2_y > J_{th,T^2_y} \Rightarrow \text{ fault in } y \text{ subspace, which is uncorrelated with } u \\
\text{otherwise} & \Rightarrow \text{ fault free.}
\end{cases}
$$

Remark 4.1. Note that the residual signal relates the input and output vectors, thus, not only the faults in the sensors (i.e., in y) and actuators (i.e., in u), but also those faults that lead to a (large) change in system parameters are detectable.

Remark 4.2. Note that, based on the identity (4.15), the follow residual vector can be constructed as

$$\mathbf{r} = \mathbf{J}_1^T \mathbf{u} - \Sigma_\kappa \mathbf{L}_1^T \mathbf{y}$$

Furthermore, based on the identities (4.12) and (4.13), the follow residual vectors can also be constructed as

$$\mathbf{r} = \mathbf{L}^T \mathbf{y} - \Sigma^T \mathbf{J}^T \mathbf{u}$$
$$\mathbf{r} = \mathbf{J}^T \mathbf{u} - \Sigma \mathbf{L}^T \mathbf{y}$$

It is easy to develop FD method in a similar way as shown in Algorithm 4.1. Without confusion, in this dissertation, the CCA-based FD methods are developed based on the residual vector (4.21).

4.3.2 An Illustrative Example

Consider an open-loop example, with a linear relationship between the input and output variables modeled as [45]

$$\mathbf{y}(k) = \Phi \mathbf{u}(k) + \mathbf{b_{obs}} + \mathbf{v}(k) \tag{4.29}$$

where $\mathbf{u} \in \mathcal{R}^l$, $\mathbf{y} \in \mathcal{R}^m$ denote the input and output variables, which are normally distributed; $\mathbf{v} \in \mathcal{R}^m$ represents the zero-mean noise, which is uncorrelated with $\mathbf{u}$ and mutually independent; and the system parameters $\mathbf{\Phi} \in \mathcal{R}^{m \times l}$ is time invariant. $\mathbf{b}_{obs} \in \mathcal{R}^m$ is a constant vector. For the simulation, let $m = 6$ and $l = 3$ to represent a six-output, three-input linear static processes.

Two thousand samples are generated to train the model. It is found that $\mathbf{L}^T \mathbf{y}$ and $\mathbf{J}^T \mathbf{u}$, have strong correlation, i.e., $\mathbf{\Sigma}_\varrho \approx \mathbf{I}_\varrho$. Hence, Q_{cca} test statistic is more appropriate. The threshold is determined at a 0.05 significance level. In addition, three fault scenarios are tested. The first fault scenario is simulated by a step change in the first actuator i.e., u_1, followed by a bias sensor in y_2 in the second fault scenario. In the last fault scenario, a parameter change is injected into the $\mathbf{\Phi}$ matrix, where $\mathbf{\Phi}(1,1)$ was changed from the nominal value to three times of the nominal one. In each scenario the change was introduced at sample 1000.

Detection results of the CCA-based method are given in Figures 4.1a, 4.1b and 4.1c. The MTFA of the proposed method is 20.02, which is acceptable. The FDR of three fault scenarios are 99.92%, 99.97% and 95.50%, respectively. The achieved results suggest that the CCA-based method can successfully detect those faults.

4.4 DCCA-based FD Method for Dynamic Processes

In this section, the CCA-based method presented in the last section is extended to deal with FD in dynamic processes operating at steady state, which is analog to the well-accepted DPCA- and DPLS-based methods [68, 123]. As introduced in Chapter 2, a standard model form of such dynamic processes is the state space representation given by (2.4)-(2.5). In our study, we further assume that the process of interest is stable. It is well known that at steady state [59], i.e., $\lim_{k \to \infty} \mu_x(k) = \mu_x$ and $\lim_{k \to \infty} \Sigma_x(k) = \Sigma_x$, where μ_x and Σ_x are constant. $\mathbf{u}(k)$ could be, in general, a constant signal (at least in mean value). Below, we propose DCCA, as an extension of CCA-based method, to detect faults in such a steady state dynamic process.

4.4.1 Modeling of Input and Output Data Sets

Based on the stochastic system model (2.4)-(2.5), the subsection investigates the dependence of future outputs ($\mathbf{y}_f$) on past inputs and outputs ($\mathbf{z}_p$) and future inputs ($\mathbf{u}_f$). To this end, we first define the data structures and sets. Suppose that s_p and s_f are the time lags.

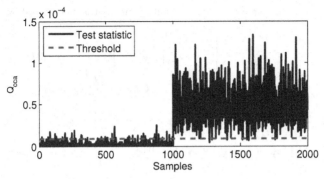

(a) Detection result for the first scenario: actuator fault

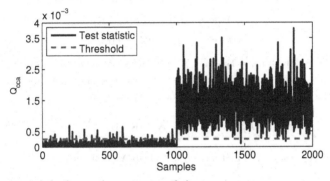

(b) Detection result for the second scenario: sensor fault

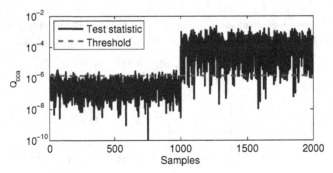

(c) Detection result for the third scenario: process fault

Figure 4.1: Detection results of the static process based on the CCA method

Let the lagged variables and the corresponding data matrices be defined by

$$
\mathbf{z}_p(k) = \begin{bmatrix} \mathbf{y}(k - s_p) \\ \cdots \\ \mathbf{y}(k - 1) \\ \mathbf{u}(k - s_p) \\ \cdots \\ \mathbf{u}(k - 1) \end{bmatrix}, \mathbf{y}_f(k) = \begin{bmatrix} \mathbf{y}(k) \\ \cdots \\ \mathbf{y}(k + s_f) \end{bmatrix}, \mathbf{u}_f(k) = \begin{bmatrix} \mathbf{u}(k) \\ \cdots \\ \mathbf{u}(k + s_f) \end{bmatrix},
$$

$$
\mathbf{Z}_p = [z_p(1), \ldots, z_p(N)] \in \mathcal{R}^{(s_p(m+l)) \times N},
$$

$$
\mathbf{Y}_f = [\mathbf{y}_f(1), \ldots, \mathbf{y}_f(N)] \in \mathcal{R}^{(s_f+1)m \times N}, \tag{4.30}
$$

$$
\mathbf{U}_f = [\mathbf{u}_f(1), \ldots, \mathbf{u}_f(N)] \in \mathcal{R}^{(s_f+1)l \times N}.
$$

It is demonstrated in [93] that representation (2.4)-(2.5) can be rewritten as

$$
\mathbf{x}(k + 1) = \mathbf{A}_K \mathbf{x}(k) + \mathbf{B}_K \mathbf{u}(k) + \mathbf{K} \mathbf{y}(k), \tag{4.31}
$$

$$
\mathbf{y}(k) = \mathbf{C} \mathbf{x}(k) + \mathbf{D} \mathbf{u}(k) + \mathbf{e}(k), \tag{4.32}
$$

where $\mathbf{A}_K = \mathbf{A} - \mathbf{KC}$, $\mathbf{B}_K = \mathbf{B} - \mathbf{KD}$, with $\mathbf{K}$ as Kalman filter gain matrix that ensures the eigenvalues of $\mathbf{A}_K$ are all located in the unit circle. $\mathbf{e}(k)$ is the innovation sequence. It is straightforward from (4.31) that the following equation holds

$$
\mathbf{x}(k) = \mathbf{A}_K^{s_p} \mathbf{x}(k - s_p) + \sum_{i=1}^{s_p} \mathbf{A}_K^{i-1} [\mathbf{K} \quad \mathbf{B}_K] \begin{bmatrix} \mathbf{y}(k - i) \\ \mathbf{u}(k - i) \end{bmatrix}. \tag{4.33}
$$

Recall that $\mathbf{A}_K$ is stable, so that a large s_p leads to $\mathbf{A}_K^{s_p} \approx 0$, then

$$
\mathbf{x}(k) \approx \mathbf{P}^{\mathrm{T}} \mathbf{z}_p(k), \tag{4.34}
$$

where $\mathbf{P}^{\mathrm{T}} = [\mathbf{P}_y \ \mathbf{P}_u]$, $\mathbf{P}_y = \begin{bmatrix} \mathbf{A}_K^{s_p-1}\mathbf{K} & \ldots & \mathbf{A}_K\mathbf{K} & \mathbf{K} \end{bmatrix}$, $\mathbf{P}_u = \begin{bmatrix} \mathbf{A}_K^{s_p-1}\mathbf{B}_K & \ldots & \mathbf{A}_K\mathbf{B}_K & \mathbf{B}_K \end{bmatrix}$. The 'past' process measurements $\mathbf{z}_p(k)$ include the process input and output data in the time period $[k - s_p, k - 1]$ as shown in (4.30). It follows from (4.31) and (4.32) that

$$
\mathbf{y}_f(k) = \mathbf{\Gamma}_{K,s_f} \mathbf{x}(k) + \mathbf{H}_{K,u,s_f} \mathbf{u}_f(k) + \mathbf{H}_{K,y,s_f} \mathbf{y}_f(k) + \mathbf{e}_f(k), \tag{4.35}
$$

where

$$
\mathbf{\Gamma}_{K,s_f} = \begin{bmatrix} \mathbf{C} \\ \mathbf{C}\mathbf{A}_K \\ \vdots \\ \mathbf{C}\mathbf{A}_K^{s_f} \end{bmatrix}, \mathbf{H}_{K,u,s_f} = \begin{bmatrix} \mathbf{D} & 0 & \ldots & 0 \\ \mathbf{C}\mathbf{B}_K & \mathbf{D} & \ddots & \vdots \\ \vdots & \ddots & \ddots & 0 \\ \mathbf{C}\mathbf{A}_K^{s_f-1}\mathbf{B}_K & \ldots & \mathbf{C}\mathbf{B}_K & \mathbf{D} \end{bmatrix},
$$

$$
\mathbf{H}_{K,y,s_f} = \begin{bmatrix} 0 & 0 & \ldots & 0 \\ \mathbf{C}\mathbf{K} & 0 & \ddots & \vdots \\ \vdots & \ddots & \ddots & 0 \\ \mathbf{C}\mathbf{A}_K^{s_f-1}\mathbf{K} & \ldots & \mathbf{C}\mathbf{K} & 0 \end{bmatrix}, \mathbf{e}_f(k) = \begin{bmatrix} \mathbf{e}(k) \\ \mathbf{e}(k + 1) \\ \vdots \\ \mathbf{e}(k + s_f) \end{bmatrix}.
$$

Based on (4.34), we have

$$(\mathbf{I} - \mathbf{H}_{K,y,s_f})\mathbf{y}_f(k) \approx \boldsymbol{\Gamma}_{K,s_f}\mathbf{P}^\mathrm{T}\mathbf{z}_p(k) + \mathbf{H}_{K,u,s_f}\mathbf{u}_f(k) + \mathbf{e}_f(k)$$

$$= [\boldsymbol{\Gamma}_{K,s_f}\mathbf{P}^\mathrm{T} \ \mathbf{H}_{K,u,s_f}]\begin{bmatrix}\mathbf{z}_p(k)\\\mathbf{u}_f(k)\end{bmatrix} + \mathbf{e}_f(k). \tag{4.36}$$

(4.36) is further written as

$$\mathbf{L}^\mathrm{T}\mathbf{y}_f(k) = \mathbf{M}^\mathrm{T}\begin{bmatrix}\mathbf{z}_p(k)\\\mathbf{u}_f(k)\end{bmatrix} + \mathbf{e}_f(k) \tag{4.37}$$

where $\mathbf{L} = (\mathbf{I} - \mathbf{H}_{K,y,s_f})^\mathrm{T}$, $\mathbf{M} = [\boldsymbol{\Gamma}_{K,s_f}\mathbf{P}^\mathrm{T} \ \mathbf{H}_{K,u,s_f}]^\mathrm{T}$.

4.4.2 DCCA-based FD Method

We now address FD in dynamic processes by applying CCA technique for residual generation. The process input and output data are constructed in a time interval, respectively denoted by $\mathbf{Y}_f$ and $\begin{bmatrix}\mathbf{Z}_p\\\mathbf{U}_f\end{bmatrix}$. Let $\begin{bmatrix}\mathbf{Z}_p\\\mathbf{U}_f\end{bmatrix}$ and $\mathbf{Y}_f$ be mean-centered, then

$$\begin{bmatrix}\boldsymbol{\Sigma}_z & \boldsymbol{\Sigma}_{z,y_f}\\\boldsymbol{\Sigma}_{y_f,z} & \boldsymbol{\Sigma}_{y_f}\end{bmatrix} \approx \frac{1}{N-1}\begin{pmatrix}\begin{bmatrix}\mathbf{Z}_p\\\mathbf{U}_f\end{bmatrix}\begin{bmatrix}\mathbf{Z}_p\\\mathbf{U}_f\end{bmatrix}^\mathrm{T} & \begin{bmatrix}\mathbf{Z}_p\\\mathbf{U}_f\end{bmatrix}\mathbf{Y}_f^\mathrm{T}\\\mathbf{Y}_f\begin{bmatrix}\mathbf{Z}_p\\\mathbf{U}_f\end{bmatrix}^\mathrm{T} & \mathbf{Y}_f\mathbf{Y}_f^\mathrm{T}\end{pmatrix}$$

By using CCA, the weighting matrices $\mathbf{J}_d$ and $\mathbf{L}_d$ are obtained from

$$\mathbf{J}_d = \boldsymbol{\Sigma}_z^{-1/2}\boldsymbol{\Gamma}(:,1:n), \quad \mathbf{L}_d = \boldsymbol{\Sigma}_{y_f}^{-1/2}\boldsymbol{\Upsilon}(:,1:n),$$
$$\boldsymbol{\Sigma}_z^{-1/2}\boldsymbol{\Sigma}_{z,y_f}\boldsymbol{\Sigma}_{y_f}^{-1/2} = \boldsymbol{\Gamma}\boldsymbol{\Sigma}\boldsymbol{\Upsilon}^\mathrm{T}, \boldsymbol{\Sigma} = \begin{bmatrix}\boldsymbol{\Sigma}_n & 0\\0 & 0\end{bmatrix} \tag{4.38}$$

where $\boldsymbol{\Sigma}_n = diag(\rho_1,\ldots,\rho_n)$. Cumulative percentage value method is a simple and effective way to determine n, which is called the order of the system [87]. Based on Theorem 4.1, it is reasonable to define a residual vector as

$$\mathbf{r}(k) = \mathbf{L}_d^\mathrm{T}\mathbf{y}_f(k) - \mathbf{M}_d^\mathrm{T}\begin{bmatrix}\mathbf{z}_p(k)\\\mathbf{u}_f(k)\end{bmatrix} \tag{4.39}$$

where $\mathbf{M}_d^\mathrm{T} = \boldsymbol{\Sigma}_n\mathbf{J}_d^\mathrm{T}$. Furthermore, similar to the static case, the covariance matrix of $\mathbf{r}(k)$ can be estimated as

$$\frac{1}{N-1}\left(\mathbf{L}_d^\mathrm{T}\mathbf{Y}_f - \mathbf{M}_d^\mathrm{T}\begin{bmatrix}\mathbf{Z}_p\\\mathbf{U}_f\end{bmatrix}\right)\left(\mathbf{L}_d^\mathrm{T}\mathbf{Y}_f - \mathbf{M}_d^\mathrm{T}\begin{bmatrix}\mathbf{Z}_p\\\mathbf{U}_f\end{bmatrix}\right)^\mathrm{T}$$
$$= \frac{\mathbf{I}_n - \boldsymbol{\Sigma}_n^2}{N-1} \tag{4.40}$$

For FD purpose, a T^2 statistic is constructed as

$$T_r^2(k) = (N-1)\mathbf{r}^{\mathrm{T}}(k)(\mathbf{I}_n - \boldsymbol{\Sigma}_n^2)^{-1}\mathbf{r}(k) \tag{4.41}$$

and the threshold J_{th,T_r^2} can be determined by

$$J_{th,T_r^2} = \chi_\alpha^2(n) \tag{4.42}$$

Denote $\mathbf{L}_{dre} = \boldsymbol{\Sigma}_{y_f}^{-1/2}\boldsymbol{\Upsilon}(:,n:(s_f+1)m-n)$ and $\mathbf{J}_{dre} = \boldsymbol{\Sigma}_z^{-1/2}\boldsymbol{\Gamma}(:,n:(s_p+s_f+1)l+s_pm-n)$, then, analog to the discussion in Section 4.3, two statistics are constructed as

$$T_z^2 = \mathbf{z}^{\mathrm{T}}\mathbf{J}_{dre}\mathbf{J}_{dre}^{\mathrm{T}}\mathbf{z} \tag{4.43}$$
$$T_{yf}^2 = \mathbf{y}_f^{\mathrm{T}}\mathbf{L}_{dre}\mathbf{L}_{dre}^{\mathrm{T}}\mathbf{y}_f \tag{4.44}$$

Hence, the corresponding thresholds are determined to be

$$J_{th,T_z^2} = \chi_\alpha^2((s_p+s_f+1)l+s_pm-n) \tag{4.45}$$
$$J_{th,T_{yf}^2} = \chi_\alpha^2((s_f+1)m-n) \tag{4.46}$$

Procedures for the proposed dynamic method are summarized in Algorithm 4.2.

Algorithm 4.2. *DCCA-based FD method*

Off-line design
S1: Center the process data to obtain $\mathbf{Z}_p$, $\mathbf{U}_f$, $\mathbf{Y}_f$
S2: Determine the number of lags s_p *and* s_f
S3: Determine the system order n *and calculate* $\mathbf{M}_d$, $\mathbf{L}_d$, $\mathbf{J}_{dre}$ *and* $\mathbf{L}_{dre}$
S4: Compute $(\mathbf{I}_n - \boldsymbol{\Sigma}_n^2)^{-1}$ *according to (4.40)*
S5: Determine the corresponding thresholds according to (4.42), (4.45) and (4.46)
On-line implementation
S6: Calculate the residual signal according to (4.39)
S7: Calculate T_r^2, T_z^2 *and* T_{yf}^2 *statistics*
S8: Make a decision based on the decision logic:

$$\begin{cases} T_r^2 > J_{th,T_r^2} & \Rightarrow \textit{fault in u and y subspaces} \\ T_r^2 \le J_{th,T_r^2} \textit{ and } T_z^2 > J_{th,T_z^2} & \Rightarrow \textit{fault in u subspace, which is uncorrelated with y} \\ T_r^2 \le J_{th,T_r^2} \textit{ and } T_{yf}^2 > J_{th,T_{yf}^2} & \Rightarrow \textit{fault in y subspace, which is uncorrelated with u} \\ \textit{otherwise} & \Rightarrow \textit{fault free.} \end{cases}$$

4.4.3 An Illustrative Example

In this subsection, the performance of the proposed DCCA-based approach is illustrated on the following dynamic system,

$$\mathbf{x}(k+1) = \mathbf{A}\mathbf{x}(k) + \mathbf{B}\mathbf{u}(k) + \boldsymbol{\eta}(k)$$
$$\mathbf{y}(k) = \mathbf{C}\mathbf{x}(k) + \boldsymbol{\varepsilon}(k)$$

where

$$\mathbf{A} = \begin{bmatrix} 0 & 1 & 0 \\ 0 & 0 & 1 \\ 0 & -0.02 & -0.4 \end{bmatrix}, \ \mathbf{B} = \begin{bmatrix} 1 & 0 \\ 0 & 1 \\ 1 & 1 \end{bmatrix}, \ \mathbf{x}_0 = \begin{bmatrix} 0 \\ 0 \\ 0 \end{bmatrix} \ \mathbf{C} = \begin{bmatrix} 1 & 1 & 0 \\ 0 & 1 & 1 \end{bmatrix}$$

$$\boldsymbol{\eta}(k) \sim \mathcal{N}(0, \operatorname{diag}(0.05, 0.05, 0.05)), \ \boldsymbol{\varepsilon}(k) \sim \mathcal{N}(0, \operatorname{diag}(0.1, 0.1))$$

For driving this system, the input variable (after centering) is generated as follows,

$$\mathbf{u}(k) \sim \mathcal{N}\left(\begin{bmatrix} 0 \\ 0 \end{bmatrix}, \begin{bmatrix} 1 & 0.1 \\ 0.1 & 2.5 \end{bmatrix} \right)$$

For off-line training, one thousand samples of fault-free input and output data are generated. To identify the parameters of DCCA-based fault detection approach, i.e., $\mathbf{L}_d$ and $\mathbf{M}_d$, we select $s_p = s_f = 6$. The threshold is determined at a 0.05 significance level.

Three fault scenarios are considered as the following:

- The first fault scenario: A step change in the first actuator, i.e.,

$$\mathbf{u}_1(k) = \begin{cases} \mathbf{u}_1(k) & k \le 400 \\ \mathbf{u}_1(k) + 1 & 400 < k \le 1000 \end{cases}$$

- The second fault scenario: A bias in the second sensor, i.e.,

$$\mathbf{y}_2(k) = \begin{cases} \mathbf{y}_2(k) & k \le 400 \\ \mathbf{y}_2(k) + 1.5 & 400 < k \le 1000 \end{cases}$$

- The third fault scenario: A process fault, i.e.,

$$\mathbf{C}(1,1) = \begin{cases} 1 & k \le 400 \\ 1.5 & 400 < k \le 1000 \end{cases}$$

Since three faults are selected in such a way that they influence the correlated input and output subspaces, we use only the T_r^2 test statistic for fault detection purpose. From Figure 4.2a, 4.2b and 4.2c, we can see that the performance in fault-free period is acceptable and all three faults are successfully detected.

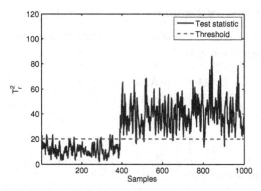

(a) Detection result for the first scenario: actuator fault

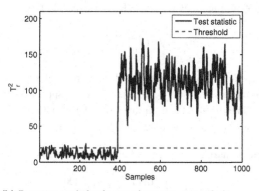

(b) Detection result for the second scenario: sensor fault

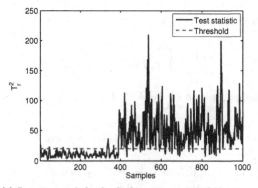

(c) Detection result for the third scenario: process fault

Figure 4.2: Detection results of the dynamic process based on DCCA method

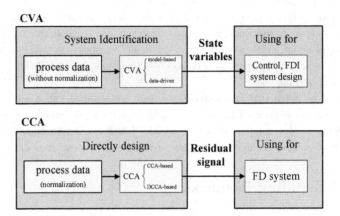

Figure 4.3: Schematic comparison between CVA-based subspace identification and the proposed methods

4.5 Discussions

Two illustrative examples have shown the effectiveness of the proposed CCA-based FD methods, which are natural extensions of the PCA- and PLS-based methods to deal with FD issues in those processes with process input and output data. Similar with DPCA and DPLS, DCCA-based FD method can only be effectively applied to LTI processes which are stable and in the steady state. It is worth noting that CCA also has been used for system identification. The early work can be found in [2]. Later on, Larimore [69] developed the canonical variate analysis (CVA)-based method for system identification, where the CCA technique is used to identify the state variables, which are also called canonical variables that represent the system dynamics. It is noted that data used in that method do not need to be mean-centered. Katayama *et al.* [62] has developed a system identification method based on the CCA technique, where data should be centered beforehand. Recently, FD methods based on CVA have received increasing attention, which are applicable for any LTI processes [87, 88, 97, 101, 111]. Different from the CVA-based methods, the core of the proposed CCA- and DCCA-based methods is residual generation without identifying the state space model. The proposed methods are mainly characterized by their significantly simplified design procedure without the system identification step. It is known that the CVA-based method for system identification is problematic when applied in closed-loop configuration. This problem comes from the projection on orthogonal complement of future input, which requires future input to be uncorrelated with past noise [109]. In other words, under the closed-loop condition, system inputs are strongly correlated with the system outputs through a feedback controller. In the proposed CCA-based method, there is no such projection. Indeed CCA constructs the residual signal by transformations of input and output. In this sense, the proposed method may be applicable to FD for the closed-loop case with a little modification, which should be explored in future work. Figure 4.3 shows a schematic comparison between CVA-based subspace identification methods and the proposed methods.

It should be noted that the proposed methods are based on canonical correlation residuals, which are generated from input variables and output variables. Hence, they are applicable to detect input faults (i.e., actuator faults) and output faults (i.e., sensor faults). Furthermore, the faults that change system parameters can also be detected.

4.6 Concluding Remarks

In this chapter, novel CCA-based FD methods have been proposed for linear static and steady dynamic processes. Compared with common MVA-based methods, e.g., PCA- and PLS-based, the proposed methods, which consider the input and output variables in both off-line training and on-line monitoring procedures, are an improvement. The core of the proposed methods is to generate residual signals by means of the CCA technique. For static processes, current input and output measurements have been used for residual generation. For dynamic processes, the process input and output data in a time interval have been used. Different from the CVA-based methods, the proposed methods are mainly characterized by their significantly simplified design procedure due to avoiding the need for system identification. The studies on numerical examples have shown the effectiveness of the proposed methods.

Note that the detection of incipient faults also is important in industry. The next chapter will focus on improving the CCA-based methods for such problems.

5 Improved CCA-based Fault Detection Methods

As discussed in subsection 1.2.2, the possible faults are divided into two categories:

- additive faults that only influence the mean value of the process;

- multiplicative faults that influence the variances, covariance, or higher-order statistical characteristics of the process.

Additive faults normally represent changes such as an abrupt increase in feed or a biased sensor, while multiplicative faults usually refer to changes, like variation in system parameters and variance of measurement noise [10, 16, 25, 90].

Compared with the large amount of research works focusing on (incipient) additive fault detection issues [47], studies on multiplicative faults are relatively few [17]. In fact, early detection of multiplicative faults is important in industry. The earlier the detection, the faster can the appropriate measures be taken to prevent accidents and serious economic losses.

It is the purpose of this chapter to improve the CCA-based methods for detecting incipient multiplicative faults in industrial processes. Specifically, we will present new methods that integrate the statistical local approach [8] into the existing CCA-based methods for detecting incipient multiplicative faults.

5.1 Background and Problem Formulation

5.1.1 Background

This subsection seeks to develop a unified framework of CCA-based FD methods for linear static and dynamic processes. In the static case, the output is assumed to be affected only by the current measurements. In the dynamic case, the process is assumed to be running in steady state and the output can depend on past measurements. Process input and output data in a time interval are used for FD.

As introduced in Chapter 4, a general canonical correlation-based residual generator can be built as

$$\mathbf{r} = \mathbf{L}^T \mathbf{y} - \mathbf{M}^T \mathbf{u} \tag{5.1}$$

Let $\boldsymbol{\pi}(k) \in \mathcal{R}^\xi$ be a data vector, and

$$\Omega_k := [\boldsymbol{\pi}(k) \ \ldots \ \boldsymbol{\pi}(k + N - 1)] \in \mathcal{R}^{\xi \times N} \tag{5.2}$$

$$\boldsymbol{\pi}_p(k) := \begin{bmatrix} \boldsymbol{\pi}(k - s_p) \\ \vdots \\ \boldsymbol{\pi}(k - 1) \end{bmatrix} \in \mathcal{R}^{s_p \xi}, \quad \boldsymbol{\pi}_f(k) := \begin{bmatrix} \boldsymbol{\pi}(k) \\ \vdots \\ \boldsymbol{\pi}(k + s_f) \end{bmatrix} \in \mathcal{R}^{(s_f+1)\xi}$$

$$\Omega_{k,p} := [\boldsymbol{\pi}_p(k) \ \ldots \ \boldsymbol{\pi}_p(k + N - 1)] = \begin{bmatrix} \Omega(k - s_p) \\ \vdots \\ \Omega(k - 1) \end{bmatrix} \in \mathcal{R}^{s_p \xi \times N} \tag{5.3}$$

$$\Omega_{k,f} := [\boldsymbol{\pi}_f(k) \ \ldots \ \boldsymbol{\pi}_f(k + N - 1)] = \begin{bmatrix} \Omega(k) \\ \vdots \\ \Omega(k + s_f) \end{bmatrix} \in \mathcal{R}^{(s_f+1)\xi \times N}$$

where s_p, s_f and N are some (large) integers; $\boldsymbol{\pi}(k)$ can be any needed vector, such as the output $\mathbf{y}(k)$ or the input $\mathbf{u}(k)$; and ξ denotes the dimension of $\mathbf{u}$ or $\mathbf{y}$. Note that, in the static case, both s_p and s_f are zero, that is, only data structure (5.2) is used. However, in the dynamic case, these two parameters may not be zero and data structure (5.3) is used.

Let the unified data structure be

$$\mathbf{u}(k) \in \mathcal{R}^{\bar{l}} := \begin{cases} \mathbf{u}(k) \in \mathcal{R}^l, & \text{static} \\ \begin{bmatrix} \mathbf{z}_p(k) \\ \mathbf{u}_f(k) \end{bmatrix} \in \mathcal{R}^{(m+l)s_p + (s_f+1)l}, & \text{dynamic} \end{cases}$$

$$\mathbf{y}(k) \in \mathcal{R}^{\bar{m}} := \begin{cases} \mathbf{y}(k) \in \mathcal{R}^m, & \text{static} \\ \mathbf{y}_f(k) \in \mathcal{R}^{(s_f+1)m}, & \text{dynamic} \end{cases}$$

Suppose that N samples of the recorded process data are collected, $\mathbf{U} \in \mathcal{R}^{\bar{l} \times N}$ and $\mathbf{Y} \in \mathcal{R}^{\bar{m} \times N}$. Let $\mathbf{U}$ and $\mathbf{Y}$ be mean centered, then

$$\begin{bmatrix} \boldsymbol{\Sigma}_u & \boldsymbol{\Sigma}_{uy} \\ \boldsymbol{\Sigma}_{uy}^{\mathrm{T}} & \boldsymbol{\Sigma}_y \end{bmatrix} \approx \frac{1}{N-1} \begin{bmatrix} \mathbf{U}\mathbf{U}^{\mathrm{T}} & \mathbf{U}\mathbf{Y}^{\mathrm{T}} \\ \mathbf{Y}\mathbf{U}^{\mathrm{T}} & \mathbf{Y}\mathbf{Y}^{\mathrm{T}} \end{bmatrix} \tag{5.4}$$

To detect faults in static and dynamic processes, Algorithms 4.1 and 4.2 can be used, respectively.

5.1.2 A Motivation Example

Consider the numerical simulation in subsection 4.3.2

$$\mathbf{y}(k) = \boldsymbol{\Phi}\mathbf{u}(k) + \mathbf{b_{obs}} + \mathbf{v}(k)$$

For the simulation, let $m = 6$ and $l = 3$ to represent a six-output, three-input linear static processes. Assume that changes occur in the system parameter $\boldsymbol{\Phi}$.

In this numerical simulation, 2000 samples are generated for training. It is found that $\mathbf{L}^T\mathbf{y}$ and $\mathbf{J}^T\mathbf{u}$, have strong correlation i.e., $\Lambda_\varrho \approx \mathbf{I}_\varrho$. Hence, Q_{cca} is used in this simulation. The threshold is determined at a significance level of 0.05. Two new data sets of 2000 samples are collected, each of which was generated by changing the value of $\Phi(3,1)$ from the nominal value to two different values, namely, 0.1 and 1.5. In each case, the parameter change was introduced at sample 1000. The first column of Figure 5.1 shows the plots of the Q_{cca} statistic for the two fault scenarios. It can be seen that the Q_{cca} statistic is not sensitive to the incipient change as shown in Figure 5.1 (middle of the first column). In general, the change is only detected in the case of a relatively large parameter variation (bottom of the first column).

From this numerical example, it is evident that the original CCA-based method fails to detect the incipient multiplicative fault. On the other hand, in practice, early detection of incipient multiplicative faults plays a significant role in preventing economic losses and hazardous accidents. Therefore, it is desirable to develop an effective method to resolve this problem.

5.2 Integrating the Statistical Local Approach into CCA for FD

In subsection 5.2.1, we briefly introduce the statistical local approach. The method to be developed will be described in the last two subsections. Based on CCA technique, subsection 5.2.2 derives the required basic residuals used in the statistical local approach, i.e., primary residual and improved residual. The FD method is proposed in subsection 5.2.3.

5.2.1 Introduction of the Statistical Local Approach

The statistical local approach is a well-established technique [8], which reduces the detection problem for multiplicative faults to the problem of detecting the mean change of a Gaussian vector. Below, we introduce the key components of this method.

Denote θ and θ_0 as parameters that represent abnormal and normal behaviors of the process. In [8], it has shown that the primary residual $\mathbf{K}(\theta_0, \mathbf{z})$ must meet the following four assumptions:

1. $E\{\mathbf{K}(\theta_0, \mathbf{z})\} = 0$ if $\theta = \theta_0$;

2. $E\{\mathbf{K}(\theta_0, \mathbf{z})\} \neq 0$ if $\theta \neq \theta_0$ but θ is in the neighborhood of θ_0;

3. $\mathbf{K}(\theta_0, \mathbf{z})$ is differentiable with respect to θ; and

4. $\mathbf{K}(\theta_0, \mathbf{z})$ exists in the vicinity of θ_0.

where the auxiliary vector z is simply composed of a finite size sample of observed measurements or the estimation error. The covariance of $K(\theta_0, z)$ is

$$R(\theta_0) \triangleq \lim_{N \to \infty} (1/N) \sum_{i=1}^{N} \sum_{j=1}^{N} E(K(\theta_0, z(i))) K(\theta_0, z(j))^T$$

Since the underlying distribution function of $K(\theta_0, z)$ is unknown and difficult to obtain, so it is difficult to directly develop fault detection algorithm based on its distributional information. Fortunately, the so-called statistical local hypothesis can overcome this problem. One of the advantage of the statistical local approach is whatever the probability distribution of the original data, it leads you to end up with a Gaussian distribution [8].

On Assumption 4, the abnormal parameter can be written as $\theta = \theta_0 + \frac{\tilde{\theta}}{\sqrt{N}}$ with $\tilde{\theta}$ is a fixed but unknown vector. The local hypothesis test is

$$
\begin{aligned}
&\text{Null hypothesis} && H_0 : \theta = \theta_0 \\
&\text{Alternative hypothesis } H_1 : \theta = \theta_0 + \frac{\tilde{\theta}}{\sqrt{N}}
\end{aligned}
\tag{5.5}
$$

Note that there exists $N \to \infty$, such that

$$\lim_{N \to \infty} \theta = \theta_0$$

In this sense, θ is in the vicinity of θ_0. $K(\theta_0, z)$ has the same covariance matrix $R(\theta_0)$ under the both hypothesis.

Given a primary residual $K(\theta_0, z)$, the improved residual is defined by

$$\phi_N \triangleq \frac{1}{\sqrt{N}} \sum_{j=1}^{N} K(\theta_0, z(j))
\tag{5.6}$$

Then the T^2 statistic is defined by

$$T_\phi^2 = \phi_N^T R(\theta_0)^{-1} \phi_N
\tag{5.7}$$

Finally, given the threshold J_{th}, the decision logic is the following:

$$
\begin{cases}
T_\phi^2 \leq J_{th}, \text{ accept the null hypothesis } H_0; \\
T_\phi^2 > J_{th}, \text{ accept the alternative hypothesis } H_1
\end{cases}
\tag{5.8}
$$

5.2.2 Derivation of the Primary Residual

As introduced in Section 4.2, $\mathbf{L}$ and $\mathbf{J}$ are consist of canonical correlation vectors. Let $\mathbf{L}_i$ and $\mathbf{J}_i$, $i = 1, 2, \ldots, \kappa$, be the ith column of the corresponding matrices, respectively.

Let $\theta_{0i} = \begin{pmatrix} \mathbf{L}_i \\ \mathbf{J}_i \end{pmatrix}$, then, based on identity (4.14) in Theorem 4.1 and the estimate (5.4), we have

$$t_i = \mathbf{U} \left(\mathbf{Y}^{\mathrm{T}} - \rho_i \mathbf{U}^{\mathrm{T}} \right) \theta_{0i}
\tag{5.9}$$

The expectation of t_i follows that

$$E\{t_i\}|_{\left(\theta_i \,=\, \theta_{0i}\right)} = 0, \; i = 1, \ldots, \kappa \qquad (5.10)$$

According to the discussion in the last subsection, t_i is defined as the *primary residual*.

Based on the CCA model, the κ pairs of $\mathbf{L}_i$ and $\mathbf{J}_i$ constitute a set of vector-valued primary residuals that are denoted as

$$\phi^{\mathrm{T}} = (t_1^{\mathrm{T}} \quad t_2^{\mathrm{T}} \quad \ldots \quad t_\kappa^{\mathrm{T}}) \qquad (5.11)$$

Without confusion, we adopt the same notation for the primary residual and the improved residual in the FD method to be developed as the ones in subsection 5.2.1.

The covariance matrix $\mathbf{S}_{\phi\phi}$ of ϕ is generally unknown but can be estimated from N observations of the statistic under the fault-free case,

$$\mathbf{S}_{\phi\phi} = \frac{1}{N-1} \sum_{i=1}^{N} \phi(i)\phi(i)^{\mathrm{T}}$$

The primary residual $\phi(\theta_0, \mathbf{z})$ meets the four assumptions described in the last subsection, where the auxiliary vector $\mathbf{z}$ here denotes the input and output data.

5.2.3 The Fault Detection Method

Based on the achieved primary residuals in subsection 5.2.2, the improved residual can be defined by

$$\phi_N = \frac{1}{\sqrt{N}} \sum_{j=1}^{N} \phi(\theta_0, \mathbf{z}(j)) \sim \mathcal{N}(0, \mathbf{S}_{\phi\phi}) \qquad (5.12)$$

then the T^2 statistic is defined by

$$T_\phi^2 = \phi_N^{\mathrm{T}} \mathbf{S}_{\phi\phi}^{-1} \phi_N \qquad (5.13)$$

Kruger et al. [66] and Hu et al. [50] have shown that the sensitivity in detecting changes in the mean value of ϕ_N decreases when N increases with newly available observation. This problem can be overcome by means of a moving window-based method. Benefits of the moving window approach have been discussed in Kano et al. [61] and Ge et al. [37]. The new improved residual is given as

$$\zeta = \frac{1}{\sqrt{w_0}} \sum_{j=k-w_0+1}^{k} \phi \qquad (5.14)$$

where w_0 is the window length. It should be noted that the determination of w_0 is done based on a tradeoff between increasing the mean time to a false alarm (or decreasing false alarms) and increasing the detectability of the faults [126]. For example, a large w_0 may reduce the number of false alarms, at the same time increasing the detection delay and, thus, reducing the sensitivity of ζ in detecting small faults. Conversely, a small window size may increase the number of false alarms, but be more sensitive to such incipient faults.

Based on the improved residual vector, the T^2 statistic is

$$T_\zeta^2 = \zeta^\mathrm{T} \mathbf{S}_{\phi\phi}^{-1} \zeta \qquad (5.15)$$

Algorithm 5.1 summaries the procedures of the proposed method for both linear static and dynamic processes. It should be noted that the difference between the static case and the dynamic case is that, in step 2, the algorithm for dynamic case should determine the number of lags s_p and s_f.

Algorithm 5.1. *Improved CCA-based FD method*

Off-line training

S1: Normalize the process data to obtain $\mathbf{U}$, $\mathbf{Y}$

S2: Determine the number of lags s_p and s_f in the dynamic case

S3: Calculate $\mathbf{L}_i, \mathbf{J}_i, \sigma_i$ using CCA

S4: Determine the window size w_0

S5: Determine the corresponding threshold J_{th,T^2} based on the χ^2 distribution

On-line monitoring based on the new measurements

S6: Compute the primary residual ϕ using (5.11)

S7: Compute the improved residual ζ using (5.14)

S8: Calculate the T_ζ^2 statistic using (5.15)

S9: Make a decision based on the decision logic:

$$\begin{cases} T_\zeta^2 > J_{th,T^2} \Rightarrow \ faulty \\ otherwise \ \Rightarrow \ fault\ free. \end{cases}$$

Improvement of the new method is, compared against the conventional one, first verified by the example described in subsection 5.1.2. The detection results using T_ζ^2 statistic are shown in the second column of Figure 5.1 where w_0 is set to 100. The fault-free case shows that the proposed method can correctly handle normal operation. From the fault cases I and II, it can be seen that the new method performs better than the conventional one. Specifically, compare the second and third rows in Figure 5.1, where the proposed method is sensitive to parameter changes, particularly incipient ones.

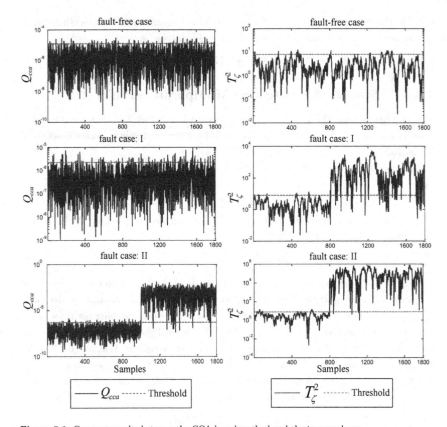

Figure 5.1: Compare results between the CCA-based method and the improved one

5.3 On Detecting Incipient Multiplicative Faults Using the T^2 Statistic

This section addresses two frequently considered questions related to detecting incipient multiplicative fault using the T^2 statistic:

1. Why is the T^2 statistic insensitive to incipient multiplicative faults?

2. Why can the statistical local approach improve its sensitivity to incipient multiplicative fault?

It is assumed that the measurement, which is affected by a multiplicative fault, is written as

$$\mathbf{y}_{mf} = \mathbf{M}\mathbf{y}$$

where $\mathbf{y}$ denotes the fault-free measurement and $\mathbf{M}$ is a square matrix representing change in variance of variables caused by the fault. The T^2 statistic is

$$T^2 = \mathbf{y}_{mf}^{\mathrm{T}} \boldsymbol{\Sigma}_y^{-1} \mathbf{y}_{mf}$$

5.3.1 T^2 Statistic for Detecting Multiplicative Faults

It is well known that the T^2 statistic in the additive fault case follows a noncentral χ^2 distribution [77, 84]. In the multiplicative fault case, the probability density function of the quadratic form of the T^2 statistic was discussed in [11].

Assume that the probability density function of the T^2 statistic in the multiplicative fault case is

$$f_{z_m}(z_m, k_m, \mathbf{M})$$

where z_m is the value of the T^2 statistic and k_m is the degrees of freedom. We define the fault detection rate in this case as

$$\mathrm{FDR} = \int_{J_{th}}^{\infty} f_{z_m}(z_m, k_m, \mathbf{M}) dz_m \tag{5.16}$$

The relationship between matrix $\mathbf{M}$ and the FDR is unclear. Figure 5.2 shows the probability density function of the T^2 statistic in the normal case and three faulty cases using the numerical example from subsection 5.1.2.[1] The corresponding matrices in the three faults are denoted as $\mathbf{M}_1$, $\mathbf{M}_2$ and $\mathbf{M}_3$. The vertical green dashed line shows the threshold J_{th}. Let S_A, S_B and S_C be the areas of the shaded regions A, B, and C. In order to visually compare the FDR for different $\mathbf{M}$ matrices, the three shaded areas (S_A, S_B and S_C) are highlighted, each of which equals $\int_0^{J_{th}} f_{z_m}(z_m, k_m, \mathbf{M}_i) dz_m$, $i = 1, 2, 3$. Since $S_A > S_B > S_C$, it is evident that the relationship between the FDR of three cases is

$$\mathrm{FDR}_{M_1} < \mathrm{FDR}_{M_2} < \mathrm{FDR}_{M_3}$$

where

$$\mathrm{FDR}_{M_1} = 1 - S_A$$
$$\mathrm{FDR}_{M_2} = 1 - S_B$$
$$\mathrm{FDR}_{M_3} = 1 - S_C$$

From this, it can be seen that the probability density function in the $\mathbf{M}_1$ case is very close to the fault-free case, so that the FDR_{M_1} is the smallest. This implies that the fault is difficult to detect using the T^2 statistic. This type of fault is considered to be an incipient multiplicative fault in this study. For different $\mathbf{M}$ matrices, as shown by $\mathbf{M}_2$ and $\mathbf{M}_3$ cases in Figure 5.2, the probability density function flattens out for larger changes in the parameters. Thus, the corresponding FDR increases, which makes detection easier.

---

[1]Matrix $\mathbf{M}$ in the normal case is the identity matrix. In faulty cases, $\mathbf{M}$ is an arbitrary matrix in $\mathcal{R}^{3\times3}$.

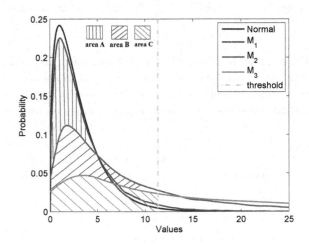

Figure 5.2: Probability density function for different multiplicative fault cases

5.3.2 Statistical Local Approach for Detecting Incipient Multiplicative Faults

The second question has been analyzed in Kruger *et al.* [66] by proving that the statistic is both sufficient in detecting changes in the eigenvectors and eigenvalues. From (5.9), it can be seen that the primary residual consists of the eigenvector θ_{0i} and eigenvalue σ_i. Thus, it is easy to prove the same result as shown in Kruger *et al.* [66]. However, a study based on checking probability density function provides an alternative way and could help us to understand the strength of the statistical local approach. This is done next.

As analyzed in Section 5.2, with the help of the statistical local approach, the multiplicative fault detection problem can be transformed into the additive fault detection problem. It is generally known that the T^2 statistic in case of an additive fault follows a noncentral χ^2 distribution [7, 122].

We assume that the probability density function of the T^2 statistic in the additive fault case is

$$\varphi_{z_a}(z_a, k_a, \delta)$$

where z_a is the value of the T^2 statistic and k_a is the degrees of freedom, δ denotes the noncentrality paramter which depends on the fault. For example, if an additive fault has the form as modeled in (3.21), then $\delta = \Xi^T \Sigma_y^{-1} \Xi f^2$ [122]. The FDR in this case is

$$\text{FDR} = \int_{J_{th}}^{\infty} \varphi_{z_a}(z_a, k_a, \delta) dz_a \tag{5.17}$$

and is an increasing function of the quantity δ [7]. This monotonic characteristic provides a clear relationship between the FDR and the fault term δ, but the one between the FDR and the fault term M in the multiplicative fault case is unclear. Figure 5.3 shows an example of the additive fault case. The plot of the probability density functions for three different δ's, where $\delta_3 > \delta_2 > \delta_1$. The vertical green dashed line denotes the threshold J_{th}. In order to visually compare the FDR for different δ's, three shaded areas are shown, each of which equals $\int_0^{J_{th}} f_{z_a}(z_a, k_a, \delta_i) dz_a$, i=1,2,3. It is evident that the relationship between the FDR for the three cases is

$$\text{FDR}_{\delta_1} < \text{FDR}_{\delta_2} < \text{FDR}_{\delta_3}$$

Since
$$\text{FDR}_{\delta_1} = 1 - S_A$$
$$\text{FDR}_{\delta_2} = 1 - S_B$$
$$\text{FDR}_{\delta_3} = 1 - S_C$$
$$S_A > S_B > S_C$$

the conclusion to be drawn here is that when the magnitude of the fault is small, that is, δ is small, it is difficult to detect the fault. Since the FDR is an increasing function of δ, the increase of δ leads to the increase of FDR as shown by δ_2 and δ_3. Motivated by this property, the statistical local approach introduced the moving window technique. Thus, the noncentrality parameter δ_w can be written as

$$\delta_w = \frac{1}{w_0} \sum_{j=1}^{w_0} \sum_{i=1}^{k_a} \bar{f}_{ji}^2$$

where $\bar{f}$ denotes the fault term in the statistical local approach. Then, δ_w is tunable by changing the window size w_0.

Therefore, when using the statistical local approach for detecting an incipient multiplicative fault, first the problem can be transformed into an additive fault problem with the help of the statistical local approach. Then, a test statistic is constructed. Its FDR is proportional to the noncentrality parameter δ_w. Finally, the FDR can be improved by increasing δ_w by tuning w_0. It should be stressed that w_0 is determined by a tradeoff between the FDR and the MTFA (or FAR).

5.4 Concluding Remarks

In this chapter, improved CCA-based FD methods have been proposed to detect incipient multiplicative faults. The improved methods have been developed based on a statistical

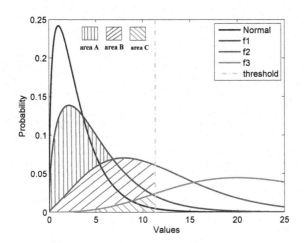

Figure 5.3: Probability density function for different additive fault cases

local approach-based residual evaluator. Although designed for both static and dynamic processes, the conventional CCA-based residual generator is only sensitive to large multiplicative faults, and not for incipient ones. The reasons why T^2 statistic is insensitive to incipient faults and how the statistical local approach improves the FDR have been further discussed. Finally, the numerical example results demonstrate that the proposed method is more powerful than the original one.

As studied in Chapter 3 and 4, the proposed CCA-based methods can only be applied for linear static processes and dynamic processes in steady state. In practice, processes are usually in dynamic with transient behavior, for example, due to fast variation in u. The next chapter will focus on FD methods for such processes in the data-driven fashion. Furthermore, the FD problem of the dynamic processes subject to deterministic disturbances will be studied.

6 A Projection-based FD method for Dynamic Processes with deterministic disturbances

Although the model-based FD methods for dynamic processes with and without deterministic disturbances are well studied, and a great number of successful implementations have been reported [40, 89, 125], the existing data-driven FD methods pay often less attention to deterministic disturbances. Recently, Luo *et al.* [75] proposed a data-driven FD approach for static processes with deterministic disturbances. The core step is to identify the maximum influence of the unknown input on the measurements. In this work, however, dynamic processes are not considered.

Motivated by these observations, in this chapter, the first objective is to generate the so-called residual signals by means of an orthogonal projection of process input data. This is considerably different from the existing model-based and data-driven FD approaches, in which residual generator is realized based on the process input and output relationship/dynamics. This way of residual generation can be done for both processes with and without deterministic disturbances. Therefore, the second objective is to deal with FD issues for dynamic processes based on the generated residual signals.

6.1 Background and Problem Formulation

6.1.1 Background

We suppose that the process under consideration can be modeled as the one described in model (2.4)-(2.5). It is further assumed that the process is stable, i.e., $\max(\text{eig}(\mathbf{A})) < 1$, where $\max(\cdot)$ denotes the maximum element in a vector. If A is unstable, the innovation model form is needed. In this chapter, the stable process will be dealt with.

For simplicity of the following study, we using the same data structure as described in (5.3).

In the model-based FD systems, it is assumed that the process model is known, namely, $\mathbf{A}$, $\mathbf{B}$, $\mathbf{C}$, $\mathbf{D}$, Σ_η, $\Sigma_{\eta\varepsilon}$, Σ_ε are known. The system model (2.4)-(2.5) can be further written as

$$\mathbf{Y} = \mathbf{\Gamma}_{s_f}X + \mathbf{H}_{u,s_f}\mathbf{U} + \mathbf{\Upsilon} \tag{6.1}$$

where $\mathbf{Y}$ and $\mathbf{U}$ consist of the temporal and past output and input data respectively, and Υ denotes the noise term; matrices Γ_{s_f} and $\mathbf{H}_{u,s_f}$ are composite of known system matrices. Note that the only unknown variable is $\mathbf{X}$.

As introduced in Section 2.4, the existing model-based residual generators, e.g., parity space- and diagnostic observer-based ones, can be generalized by the kernel representation-based residual generator, which is described as

$$\mathbf{r}(z) = \mathcal{K} \begin{bmatrix} \mathbf{u}(z) \\ \mathbf{y}(z) \end{bmatrix}$$

where z denotes the z-transformation operator. Since the process model is not always available, the kernel representation $\mathcal{K}$ should be first identified before implementation. There are a great number of successful applications [24, 28, 30]. For example, let Ψ represents the data-driven realization of the kernel representation [24]. The solution Ψ allows a residual generation in the form of

$$\mathbf{r}(k) = \Psi \begin{bmatrix} \mathbf{u}_s(k) \\ \mathbf{y}_s(k) \end{bmatrix} \tag{6.2}$$

where $\mathbf{y}_s(k)$, $\mathbf{u}_s(k)$ denote the process input and output data in the time interval $[k, k+s_f]$.

6.1.2 Problem Formulation

As can be seen in Figures 6.1a and 6.1b, the residual generation of existing model-based and data-driven FD methods is realized based on the process input and output relationship/dynamics. For example, the data-driven FD methods seek to identify the kernel representation Ψ of the process. In this chapter, we propose an alternative way to generate the residuals through a projection of process input data, as shown in Figure 6.1c. It can be seen that the residual-like signal is a matrix, which will be explained in Section 6.3. Based on the achieved residual signals, the fault detection issue for dynamic process with deterministic disturbances will be studied.

6.2 I/O Data Model with Deterministic Disturbances and Faults

The system model (2.4)-(2.5) is extended to include disturbances and faults

$$\mathbf{x}(k+1) = \mathbf{A}\mathbf{x}(k) + \mathbf{B}\mathbf{u}(k) + \mathbf{E}_d\mathbf{d}(k) + \mathbf{E}_f\mathbf{f}(k) + \boldsymbol{\eta}(k) \tag{6.3}$$

$$\mathbf{y}(k) = \mathbf{C}\mathbf{x}(k) + \mathbf{D}\mathbf{u}(k) + \mathbf{F}_d\mathbf{d}(k) + \mathbf{F}_f\mathbf{f}(k) + \boldsymbol{\varepsilon}(k) \tag{6.4}$$

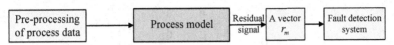

(a) The model-based FD system

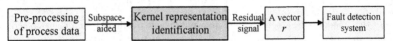

(b) The data-driven FD system

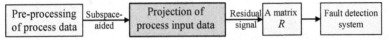

(c) An alternative data-driven FD system

Figure 6.1: Schematic comparison between existing methods and the proposed method

where $\mathbf{f}(k) \in \mathcal{R}^{k_f}$ represents all possible faults; $\mathbf{d}(k) \in \mathcal{R}^{k_d}$ represents $l_{2,[k,k+N]}$-bounded disturbance:

$$\sum_{i=0}^{N} \mathbf{d}^2(k+i) \leq \delta_d^2(N) \tag{6.5}$$

with N denoting the length of the evaluation window and δ_d is a unknown constant as boundness.

As introduced in subsection 2.1.2, a sensor fault is modeled by setting $\mathbf{E}_f = 0$, $\mathbf{F}_f = \mathbf{I}_{k_f}$; an actuator fault is often formulated by setting $\mathbf{E}_f = \mathbf{B}$, $\mathbf{F}_f = \mathbf{D}$; while a process fault can be modeled by $\mathbf{E}_f = \mathbf{E}_p$, $\mathbf{F}_f = \mathbf{F}_p$ for some $\mathbf{E}_p$ and $\mathbf{F}_p$.

Adopting the same data structure as described in (5.3), the system model (6.3)-(6.4) can be further written into

$$\begin{aligned}
\mathbf{y}_s(k) &= \mathbf{\Gamma}_{s_f}\mathbf{x}(k) + \mathbf{H}_{u,s_f}\mathbf{u}_s(k) + \mathbf{H}_{d,s_f}\mathbf{d}_s(k) \\
&\quad + \mathbf{H}_{f,s_f}\mathbf{f}_s(k) + \mathbf{H}_{\eta,s_f}\mathbf{\eta}_s(k) + \mathbf{\varepsilon}_s(k) \in \mathcal{R}^{(s_f+1)m\times 1}
\end{aligned} \tag{6.6}$$

where

$$\mathbf{\Gamma}_{s_f} = \begin{bmatrix} \mathbf{C} \\ \mathbf{CA} \\ \vdots \\ \mathbf{CA}^{s_f} \end{bmatrix} \in \mathcal{R}^{(s_f+1)m\times n}, \quad \mathbf{H}_{u,s_f} = \begin{bmatrix} \mathbf{D} & 0 & \cdots & 0 \\ \mathbf{CB} & \mathbf{D} & \ddots & \vdots \\ \vdots & \ddots & \ddots & 0 \\ \mathbf{CA}^{s_f-1}\mathbf{B} & \cdots & \mathbf{CB} & \mathbf{D} \end{bmatrix}$$

$$\mathbf{H}_{d,s_f} = \begin{bmatrix} \mathbf{F}_d & 0 & \cdots & 0 \\ \mathbf{CE}_d & \mathbf{F}_d & \ddots & \vdots \\ \vdots & \ddots & \ddots & 0 \\ \mathbf{CA}^{s_f-1}\mathbf{E}_d & \cdots & \mathbf{CE}_d & \mathbf{F}_d \end{bmatrix}, \quad \mathbf{H}_{f,s_f} = \begin{bmatrix} \mathbf{F}_f & 0 & \cdots & 0 \\ \mathbf{CE}_f & \mathbf{F}_f & \ddots & \vdots \\ \vdots & \ddots & \ddots & 0 \\ \mathbf{CA}^{s_f-1}\mathbf{E}_f & \cdots & \mathbf{CE}_f & \mathbf{F}_f \end{bmatrix}$$

$$\mathbf{H}_{\eta,s_f} = \begin{bmatrix} \mathbf{0} & \mathbf{0} & \dots & \mathbf{0} \\ \mathbf{C} & \mathbf{0} & \ddots & \vdots \\ \vdots & \ddots & \ddots & \mathbf{0} \\ \mathbf{CA}^{s_f-1} & \dots & \mathbf{C} & \mathbf{0} \end{bmatrix}$$

Note that model (6.6) describes the input and output relationship in dependence on the past state variable $\mathbf{x}(k)$. Furthermore, based on model (6.3)-(6.4), we have

$$\mathbf{x}(k) = \mathbf{A}^{s_p}\mathbf{x}(k - s_p) + \sum_{i=1}^{s_p} \mathbf{A}^{i-1}\mathbf{B}\mathbf{u}(k - i) + \sum_{i=1}^{s_p} \mathbf{A}^{i-1}\mathbf{E}_f\mathbf{f}(k - i)$$
$$+ \sum_{i=1}^{s_p} \mathbf{A}^{i-1}\mathbf{E}_d\mathbf{d}(k - i) + \sum_{i=1}^{s_p} \mathbf{A}^{i-1}\boldsymbol{\eta}(k - i) \tag{6.7}$$

Substituting (6.7) into (6.6), gives

$$\mathbf{y}_s(k) = \boldsymbol{\Gamma}_{s_f}[\mathbf{A}^{s_p}\mathbf{x}(k - s_p) + \mathbf{P}_u^{\mathrm{T}}\mathbf{u}_p(k) + \mathbf{P}_f^{\mathrm{T}}\mathbf{f}_p(k)) + \mathbf{P}_d^{\mathrm{T}}\mathbf{d}_p(k) + \mathbf{P}_{\eta}^{\mathrm{T}}\boldsymbol{\eta}_p(k)]$$
$$+ \mathbf{H}_{u,s_f}\mathbf{u}_s(k) + \mathbf{H}_{d,s_f}\mathbf{d}_s(k) + \mathbf{H}_{f,s_f}\mathbf{f}_s(k) + \mathbf{H}_{\eta,s_f}\boldsymbol{\eta}_s(k) + \boldsymbol{\varepsilon}_s(k) \tag{6.8}$$

where $\mathbf{P}_u^{\mathrm{T}} = \begin{bmatrix} \mathbf{A}^{s_p-1}\mathbf{B} & \dots & \mathbf{AB} & \mathbf{B} \end{bmatrix}$, $\mathbf{P}_d^{\mathrm{T}} = \begin{bmatrix} \mathbf{A}^{s_p-1}\mathbf{E}_d & \dots & \mathbf{AE}_d & \mathbf{E}_d \end{bmatrix}$, $\mathbf{P}_f^{\mathrm{T}} = \begin{bmatrix} \mathbf{A}^{s_p-1}\mathbf{E}_f & \dots & \mathbf{AE}_f & \mathbf{E}_f \end{bmatrix}$, $\mathbf{P}_{\eta}^{\mathrm{T}} = \begin{bmatrix} \mathbf{A}^{s_p-1} & \dots & \mathbf{A} & \mathbf{I}_n \end{bmatrix}$. The past process measurements $\mathbf{u}_p(k)$ include the process input data in the time period $[k - s_p, k - 1]$.

Since $\max \mathrm{eig}(\mathbf{A}) < 1$, it follows that $\lim_{s_p \to \infty} \mathbf{A}^{s_p} = 0$. Hence, the term $\mathbf{A}^{s_p}\mathbf{x}(k - s_p)$ is negligible when s_p is a large integer. As a result, (6.8) can be rewritten into

$$\mathbf{y}_s(k) \approx \mathbf{H}_{u,s_p,s_f}\mathbf{u}_{p,s}(k) + \mathbf{H}_{d,s_p,s_f}\mathbf{d}_{p,s}(k) + \mathbf{H}_{e,s_p,s_f}\mathbf{e}_{p,s}(k) + \mathbf{H}_{f,s_p,s_f}\mathbf{f}_{p,s}(k) \tag{6.9}$$

where

$$\mathbf{H}_{u,s_p,s_f} = \begin{bmatrix} \boldsymbol{\Gamma}_{s_f}\mathbf{P}_u^{\mathrm{T}} & \mathbf{H}_{u,s_f} \end{bmatrix}, \mathbf{u}_{p,s}(k) = \begin{bmatrix} \mathbf{u}_p(k) \\ \mathbf{u}_s(k) \end{bmatrix}$$

$$\mathbf{H}_{d,s_p,s_f} = \begin{bmatrix} \boldsymbol{\Gamma}_{s_f}\mathbf{P}_d^{\mathrm{T}} & \mathbf{H}_{d,s_f} \end{bmatrix}, \mathbf{d}_{p,s}(k) = \begin{bmatrix} \mathbf{d}_p(k) \\ \mathbf{d}_s(k) \end{bmatrix}$$

$$\mathbf{H}_{f,s_p,s_f} = \begin{bmatrix} \boldsymbol{\Gamma}_{s_f}\mathbf{P}_f^{\mathrm{T}} & \mathbf{H}_{f,s_f} \end{bmatrix}, \mathbf{f}_{p,s}(k) = \begin{bmatrix} \mathbf{f}_p(k) \\ \mathbf{f}_s(k) \end{bmatrix}$$

$$\mathbf{H}_{e,s_p,s_f} = \begin{bmatrix} \boldsymbol{\Gamma}_{s_f}\mathbf{P}_w^{\mathrm{T}} & \mathbf{H}_{\eta,s_f} & \mathbf{I}_{(s_f+1)m} \end{bmatrix}, \mathbf{e}_{p,s}(k) = \begin{bmatrix} \boldsymbol{\eta}_p(k) \\ \boldsymbol{\eta}_s(k) \\ \boldsymbol{\varepsilon}_s(k) \end{bmatrix}$$

Collect N samples and formulate the input/output model as (5.3), we have

$$\mathbf{Y}_{k,s} = \mathbf{H}_{u,s_p,s_f}\mathbf{U}_{k,p,s} + \mathbf{H}_{d,s_p,s_f}\mathbf{D}_{k,p,s} + \mathbf{H}_{e,s_p,s_f}\mathbf{E}_{k,p,s} + \mathbf{H}_{f,s_p,s_f}\mathbf{F}_{k,p,s} \tag{6.10}$$

6.3 The Projection-based Data-Driven FD Method

The basic idea of the proposed method consists of two major parts. The first part is to generate a residual-like signal, which is independent of $\mathbf{u}(k)$. Then, the second part is to construct an evaluation function, threshold setting as well as the decision making. In order to develop FD method in fault-free data, fault term $\mathbf{F}_{k,p,s}$ in (6.10) is ignored.

For the first part, let the orthogonal projection be

$$\Pi_{\mathbf{U}_{k,p,s}}^{\perp} = \mathbf{I} - \mathbf{U}_{k,p,s}^{\mathrm{T}}(\mathbf{U}_{k,p,s}\mathbf{U}_{k,p,s}^{\mathrm{T}})^{-1}\mathbf{U}_{k,p,s} \tag{6.11}$$

then eliminating $\mathbf{U}_{k,p,s}$ by post-multiplying $\Pi_{\mathbf{U}_{k,p,s}}^{\perp}$ on both sides of (6.10)

$$\begin{aligned}
\mathbf{Y}_{k,s}\Pi_{\mathbf{U}_{k,p,s}}^{\perp} &= \mathbf{H}_{u,s_p,s_f}\mathbf{U}_{k,p,s}\Pi_{\mathbf{U}_{k,p,s}}^{\perp} + \mathbf{H}_{d,s_p,s_f}\mathbf{D}_{k,p,s}\Pi_{\mathbf{U}_{k,p,s}}^{\perp} \\
&\quad + \mathbf{H}_{e,s_p,s_f}\mathbf{E}_{k,p,s}\Pi_{\mathbf{U}_{k,p,s}}^{\perp} \\
&= \mathbf{H}_{d,s_p,s_f}\mathbf{D}_{k,p,s}\Pi_{\mathbf{U}_{k,p,s}}^{\perp} + \mathbf{H}_{e,s_p,s_f}\mathbf{E}_{k,p,s} \tag{6.12}
\end{aligned}$$

Note that $\mathbf{E}_{k,p,s}\Pi_{\mathbf{U}_{k,p,s}}^{\perp} = \mathbf{E}_{k,p,s}$ since $\mathbf{e}_{p,s}(k)$ and $\mathbf{u}_{p,s}(k)$ are independent. The achieved term $\mathbf{Y}_{k,s}\Pi_{\mathbf{U}_{k,p,s}}^{\perp}$ is the residual signal, which is affected by the deterministic disturbance and noise. Note that, in the conventional residual generation-based methods [24], the constructed residual is a vector. Hence, a parameter identification step is necessary. In our case, the residual signal is generated by an orthogonal projection of input data. Therefore, the need for parameter identification is obviated, and more, the achieved evaluation function is a matrix.

For fault detection, the influence of the noise can be included or removed by setting a threshold. In [23], the case with threshold setting for processes with noise and deterministic disturbances has been studied. Therefore, in this study, the case with noise elimination is investigated.

In order to remove the influence of the noise, an instrument variable $\mathbf{Z}_k \in \mathcal{R}^{\zeta \times N}$ can be introduced, in which ζ represents any suitable dimension. Assuming $\mathbf{Z}$ is statistically independent of $\mathbf{E}$, we have

$$\begin{aligned}
\frac{\mathbf{Y}_{k,s}\Pi_{\mathbf{U}_{k,p,s}}^{\perp}\mathbf{Z}_k^{\mathrm{T}}}{N} &= \frac{\mathbf{H}_{d,s_p,s_f}\mathbf{D}_{k,p,s}\Pi_{\mathbf{U}_{k,p,s}}^{\perp}\mathbf{Z}_k^{\mathrm{T}}}{N} + \frac{\mathbf{H}_{e,s_p,s_f}\mathbf{E}_{k,p,s}\mathbf{Z}_k^{\mathrm{T}}}{N} \\
&= \frac{\mathbf{H}_{d,s_p,s_f}\mathbf{D}_{k,p,s}\Pi_{\mathbf{U}_{k,p,s}}^{\perp}\mathbf{Z}_k^{\mathrm{T}}}{N} \tag{6.13}
\end{aligned}$$

where N is used to normalize it. (6.13) is a natural result through the projection as is often done in subspace system identification literature without considering disturbances. The approximation part is achieved using an instrument variable method, in which the instrument variable $\mathbf{Z}_k$ is chosen as uncorrelated with the noise term $\mathbf{e}_{p,s}(k)$.

Define a symmetric matrix

$$\mathbf{R} = \frac{\mathbf{Y}_{k,s}\Pi_{\mathbf{U}_{k,p,s}}^{\perp}\mathbf{Z}_k^{\mathrm{T}}}{N}\left(\frac{\mathbf{Y}_{k,s}\Pi_{\mathbf{U}_{k,p,s}}^{\perp}\mathbf{Z}_k^{\mathrm{T}}}{N}\right)^{\mathrm{T}} = \mathbf{H}_{d,s_p,s_f}\tilde{\mathbf{D}}_{k,p,s}\tilde{\mathbf{D}}_{k,p,s}^{\mathrm{T}}\mathbf{H}_{d,s_p,s_f}^{\mathrm{T}} \tag{6.14}$$

where $\tilde{\mathbf{D}}_{k,p,s} = \frac{\mathbf{D}_{k,p,s}\Pi_{\tilde{\mathbf{U}}_{k,p,s}}^{\perp}\mathbf{Z}_{k}^{\mathrm{T}}}{N}$, $\mathrm{rank}(\mathbf{R}) \leq \min((s_f + 1)m, \zeta)$. Therefore, to achieve a successful fault detection, we focus on monitoring changes in matrix $\mathbf{R}$. Testing a change of matrix turns out to be very complicated and not well developed, since there are difficulties in the determination of the confidence domain for a non-constant matrix [42]. In practice, a scalar measure is often used to evaluate a matrix. Since the process model is unknown, we construct an evaluation function

$$J = \mathrm{tr}(\mathbf{H}_{d,s_p,s_f}\tilde{\mathbf{D}}_{k,p,s}\tilde{\mathbf{D}}_{k,p,s}^{\mathrm{T}}\mathbf{H}_{d,s_p,s_f}^{\mathrm{T}}) = \mathrm{tr}(\tilde{\mathbf{D}}_{k,p,s}^{\mathrm{T}}\mathbf{H}_{d,s_p,s_f}^{\mathrm{T}}\mathbf{H}_{d,s_p,s_f}\tilde{\mathbf{D}}_{k,p,s}) \tag{6.15}$$

to evaluate the matrix $\mathbf{R}$. Under the bounded condition (6.5), it is reasonable to assume that $\mathrm{tr}(R)$ is bounded by a unknown constant and define it as the threshold J_{th}. Analytically setting the threshold is intractable due to the unknown the confidence domain of J. Provided that the training data are sufficiently large, the threshold can be estimated by off-line analysis.

In the fault-free process running, it is expected that $J < J_{th}$. If there is presence of a fault, based on (6.10) and (6.13), it is evident that

$$J = \mathrm{tr}(\tilde{\mathbf{D}}_{k,p,s}^{\mathrm{T}}\mathbf{H}_{d,s_p,s_f}^{\mathrm{T}}\mathbf{H}_{d,s_p,s_f}\tilde{\mathbf{D}}_{k,p,s} + \tilde{\mathbf{F}}_{k,p,s}^{\mathrm{T}}\mathbf{H}_{f,s_p,s_f}^{\mathrm{T}}\mathbf{H}_{f,s_p,s_f}\tilde{\mathbf{F}}_{k,p,s})$$
$$= \mathrm{tr}(\tilde{\mathbf{D}}_{k,p,s}^{\mathrm{T}}\mathbf{H}_{d,s_p,s_f}^{\mathrm{T}}\mathbf{H}_{d,s_p,s_f}\tilde{\mathbf{D}}_{k,p,s}) + \mathrm{tr}(\tilde{\mathbf{F}}_{k,p,s}^{\mathrm{T}}\mathbf{H}_{f,s_p,s_f}^{\mathrm{T}}\mathbf{H}_{f,s_p,s_f}\tilde{\mathbf{F}}_{k,p,s}) \tag{6.16}$$

with $\tilde{\mathbf{F}}_{k,p,s} = \frac{\mathbf{F}_{k,p,s}\Pi_{\tilde{\mathbf{U}}_{k,p,s}}^{\perp}\mathbf{Z}_{k}^{\mathrm{T}}}{N}$, which tells us that if the fault term $\mathbf{H}_{f,s_p,s_f}\tilde{\mathbf{F}}_{k,p,s}$ is not zero, then $J > J_{th}$ is ensured. Thus the fault is detectable.

6.3.1 On-line Update Orthogonal Projection Matrix

In order to on-line realize the proposed FD method, it is desirable to recursively compute the orthogonal projection matrix $\Pi_{\mathbf{U}_{k,p,s}}^{\perp}$. Recall the orthogonal projection matrix $\Pi_{\mathbf{U}_{k,p,s}}^{\perp}$ only involves the computation of the pseudo-inverse. As $\mathbf{U}_{k,p,s} \in \mathcal{R}^{(s_p+s_f+1)l \times N}$ is a rectangular matrix with $N > (s_p + s_f + 1)l$ and its pseudo-inverse $\mathbf{W}_1 = (\mathbf{U}_{k,p,s}\mathbf{U}_{k,p,s}^{\mathrm{T}})^{-1}\mathbf{U}_{k,p,s}$. When a new sample (column vector) is available, an updated matrix $\mathbf{U}_{k,p,s}$ is formulated by appending the fresh column and deleting the first column of $\mathbf{U}_{k,p,s}$, which gives

$$\mathbf{U}_{k,p,s} = [c_1, c_2, \ldots, c_N] \tag{6.17}$$
$$\mathbf{U}_{p,s,u} = [c_2, c_3, \ldots, c_N, c_{N+1}] \tag{6.18}$$

Note that $\mathbf{U}_{k,p,s}$ and $\mathbf{U}_{p,s,u}$ contain $N - 1$ identical columns. It is required to compute

$$\mathbf{W}_2 = (\mathbf{U}_{p,s,u}\mathbf{U}_{p,s,u}^{\mathrm{T}})^{-1}\mathbf{U}_{p,s,u}$$

using W_1 and $\mathbf{u}_{p,s}$, without using any inversion subroutines. Details for calculation of $\mathbf{W}_2$ can be found in [4, 57].

Finally, the orthogonal projection matrix is updated as

$$\Pi_{\mathbf{U}_{p,s,u}}^{\perp} = \mathbf{I} - \mathbf{U}_{p,s,u}^{\mathrm{T}}\mathbf{W}_2 \tag{6.19}$$

6.3.2 Algorithm for the Proposed Approach

The algorithm of the proposed method consists of two parts, namely off-line design and on-line implementation. In the first part, the primary step is to use the collected fault-free data to construct the residual signal. Subsequently, the evaluation function is established. Then, the threshold J_{th} can be estimated based on the training data. In the second part, the on-line measurements are used to update the orthogonal matrix, then the fault detection decision is made based on comparison of the calculated evaluation function and the threshold. For better understanding, the algorithm of data-driven FD is summarized as Algorithm 6.1

Algorithm 6.1.

Off-line design

S 1: Collect data sets $\boldsymbol{y}_s$, $\boldsymbol{u}_{p,s}$ and $\boldsymbol{Z}$, and construct $\frac{1}{N} \boldsymbol{Y}_{k,s} \Pi_{U_{k,p,s}^{\perp}} \boldsymbol{Z}_k^T$.

S 2: Compute the evaluation function J based on (6.15).

S 3: Determine the threshold J_{th} based on training data

On-line implementation

S 4: Stack the on-line data $\boldsymbol{y}_s(k)$ and $\boldsymbol{u}_{p,s}(k)$.

S 5: Recursive update the orthogonal projection matrix based on (6.19).

S 6: Construct $\boldsymbol{Y}_{k,s}$ and $\boldsymbol{Z}_k$ by adding a new sample and moving the first column analog to (6.18).

S 7: Build the evaluation function s_r.

S 8: Check the decision logic:

$$\begin{cases} J > J_{th} \Rightarrow \ faulty, \\ J \leq J_{th} \Rightarrow \ fault\ free. \end{cases}$$

6.3.3 Discussions

The major difference between the proposed method, the existing model-based, and data-driven FD methods is the way of residual generation. In the proposed method, no first principles models are required and the residual signals are generated by means of an orthogonal projection of process input data. Compared with the existing data-driven FD methods, the proposed method avoids the parameter identification procedure. By recursively updating the projection matrix, it allows one to solve the fault detection problem in real time. It is clear that the number of evaluation window N determines the computational complexity of the proposed methods, which also affects the fault detection performance. Since there is a compromise on computational complexity and the fault detection performance, N should be designed based on the empirical criterion.

For practical applications, several limitations of the proposed method should be pointed out. As mentioned previously, the proposed approach is applicable to detect all three types of faults: sensor faults, actuator faults and process faults. However, in the following cases, the proposed method is not always applicable. Firstly, if we consider the fault term as described in (6.16), a necessary condition for fault detectability is

$$\mathbf{F}_{k,p,s}\Pi^{\perp}_{\mathbf{U}_{k,p,s}}\mathbf{Z}^{\mathrm{T}}_k \neq 0$$

For example, the proposed method is not always applicable for actuator fault, if it is proportional to the input signals. Secondly, the orthogonal projection may hide information about disturbance due to $\mathrm{rank}(\Pi^{\perp}_{\mathbf{U}_{k,p,s}}) \leq N - (s_p + s_f + 1)l$.

6.4 Concluding Remarks

In this chapter, an alternative data-driven FD method for dynamic processes with deterministic disturbances has been proposed. Different from the existing model-based and data-driven FD approaches, the proposed method generates the residual signal, which is in the form of a matrix, by means of an orthogonal projection of the process input data. This way of residual generation, on one hand, avoids the parameter identification step; on the other hand, it allows us to deal with the deterministic disturbances. Finally, it has shown that the recursively updating the projection matrix makes the proposed method on-line available.

7 Benchmark Studies

This chapter demonstrates the applications of the proposed methods in Chapter 4, 5 and 6 to benchmark studies. Depending on the application scope of different methods, four benchmarks are used. The methods proposed in Chapter 4, including the static CCA and dynamic CCA, are applied to the alumina evaporation process (AEP) [15]. To illustrate the effectiveness of the improved CCA methods [16], a laboratory setup of continuous stirred tank heater (CSTH) is used for the static case, and the Tennessee Eastman (TE) benchmark process is adopted for the dynamic case. Finally, the method proposed in Chapter 6 is applied to an inverted pendulum system.

7.1 Case Study on the Alumina Evaporation Benchmark Process

7.1.1 Process Description

A typical alumina evaporation process consists of serially connected units, including film tube evaporators, a forced-circulation evaporator (EVA), preheaters, flash evaporators, water collectors, and a condenser. Figure 7.1 shows the flow sheet of a four-effect falling film alumina evaporation process, where the four falling film tube evaporators (E1–E4) are connected in series. The required heat is supplied to the first evaporator by live steam. The vapor produced by each of the first three evaporators, is used as the heating source for the subsequent evaporators in series. Then, the vapor produced by the fourth evaporator is condensed in condenser and discharged. Meanwhile, at another site, the feed is injected into the system at the third and the fourth evaporators, and flows backward to vapor. Between two adjacent evaporators, there are preheaters (PH1–PH3), which are used to preheat the solution fed into the previous evaporators. In the preheater, the heating source comes from the vapor produced by previous evaporators and flash evaporators. Finally, the solution leaves from the first evaporator and goes through the flash evaporators (FT1–FT3), where the final product is discharged.

According to our knowledge about this process, 35 measured variables are collected in Table 7.1 , including temperature, pressure and flow in the whole process. In Table 7.2, the eight input variables are product flow rates of each of the evaporators and flash evaporators, as well as fresh steam flow rate. It should be noted that the output flow of the

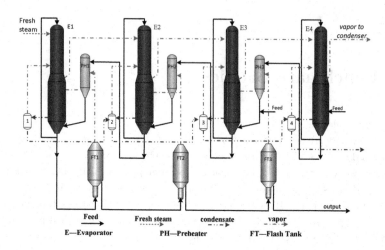

Figure 7.1: Flow sheet of four effect falling film alumina evaporation process

4th evaporator, output flow of the first flash evaporator and output flow of the second flash evaporator are the corresponding manipulated variable 6, 7 and 8, respectively. The final product concentration of three components, temperature and pressure of outlet liquor and vapor from each of the evaporators and flash evaporators present as 27 output variables.

To accurately simulate this process, a simulator has been established based on the embedded first principles model [14, 112]. The simulator has been verified by the plant measurements, and can mimic the commonly found faults. Therefore, it will be used for our application study.

7.1.2 Application of the CCA-based Methods

In this subsection, we will apply Algorithm 4.1 and 4.2 to detect faults in AEP. Meanwhile, to guide the process operator selecting a suitable method, the performance of the proposed FD methods will be evaluated in terms of MTFA, FAR and FDR. A high score method is viewed as the one has lower FAR or larger MTFA, and higher FDR.

For the study, the sampling interval is set to two minutes and simulation runs for 33 h, with $N = 990$ samples for each scenario. The training data with 500 samples are collected under the fault-free case. Table 7.3 shows the parameters determined in both the static and dynamic methods. $s_f = s_p$ is predefined, and the process order n is determined using the method in [88]. The significance level α for determine the threshold is set to be 0.05.

To show the fault detection performance of the proposed methods in this chapter, we simulates four typical fault scenarios that happening during the alumina evaporation

Table 7.1: Process variables in AEP

Block	Description	Variable
Raw liquor	Raw liquor Temp.	T_raw
	Raw liquor Flow	F_raw
Fresh steam	Fresh steam Temp.	T_fs
	Fresh steam Flow	F_fs
Evaporator i	ith Evaporator Liquor Inlet Temp.	T_Ei_Lin
	ith Evaporator Liquor Inlet Flow	F_Ei_Lin
	ith Evaporator Liquor Outlet Temp.	T_Ei_Lout
	ith Evaporator Liquor Outlet Flow	F_Ei_Lout
	ith Evaporator Vapor Outlet Temp.	T_Ei_Vout
	ith Evaporator Vapor Outlet Flow	F_Ei_Vout
	ith Evaporator Vapor Outlet Pre.	P_Ei_Vout
	$i = 1, 2, 3, 4$	
Flash EVA. j	jth Flash Evaporator Vapor Outlet Temp.	T_FEj_Vout
	jth Flash Evaporator Vapor Outlet Pre.	P_FEj_Vout
	jth Flash Evaporator Liquor Outlet Temp.	T_FEj_Lout
	jth Flash Evaporator Liquor Outlet Flow	F_FEj_Lout
	$j = 1, 2, 3$	

Table 7.2: Manipulated variables in AEP

No.	Description	Variable
1	Fresh steam Flow	F_sf
2	the fourth Evaporator Liquor Inlet Flow	F_E4_Lin
3	the third Evaporator Liquor Inlet Flow	F_E3_Lin
4	the second Evaporator Liquor Inlet Flow	F_E2_Lin
5	the first Evaporator Liquor Inlet Flow	F_E1_Lin
6	the first Flash Evaporator Liquor Inlet Flow	F_FE1_Fin
7	the second Flash Evaporator Liquor Inlet Flow	F_FE2_Fin
8	the third Flash Evaporator Liquor Inlet Flow	F_FE3_Fin

process. The faults are introduced into each run at the sample 400 until the end of simulation and listed as follows:

- Outlet sensor bias: Sensor faults are quite common in practice, among which the most common one is offset bias. We simulate a bias of 50 m^3/h in the outlet sensor of the first evaporator.

Table 7.3: Parameters used for CCA-based methods

Method	m	l	s_f	s_p	κ/n
static method (CCA)	27	8	-	-	5
dynamic method (DCCA)	27	8	3	3	62

Table 7.4: FDR for fault scenarios in AEP (%)

Method	#1	#2	#3	#4
static method (CCA)	62.44	80.70	90.69	87.01
dynamic method (DCCA)	99.64	89.17	98.65	89.02

- Scaling in the evaporator: In practice, water vapor attached to the outside and impurities deposits gradually on the inside of pipe. As a result, the heat transfer efficiency will be reduced. Here we use a ramp function (slope parameter is 10^{-3}) starting from the sample 400 to simulate a scaling fault in the fourth evaporator.

- Vapor inlet valve sticking: Valves are the most widely used actuators in the chemical industry. They are frequently subject to malfunctions. Among them, sticking is a very common one. The third fault simulates that the vapor inlet valve of fourth evaporator is sticking at 40%.

- Leakage in the valve: such a fault happen more in the heavily used components. It can, not only increase the operation cost, but also may cause (serious) environmental pollution. The fourth fault is simulated by a leakage in the fresh steam valve, represented by a 10% decrease in the fresh steam flow.

Table 7.4 offers comparison results for FDR of the two proposed FD methods, where #i denotes the ith fault scenario. For the first fault scenario, Figure 7.2 shows the CCA-based detection results. As can be seen, the method can detect this fault, but, with a low FDR. DCCA-based detection result is shown in Figure 7.3, where the FDR has been significantly improved. Additionally, in fault-free period, DCCA-based method has large MTFA (or generates fewer false alarms), which demonstrates its performance against noise. For the second fault scenario, Figures 7.4 and 7.5 show that both CCA-based and DCCA-based method can detect this fault. However, the CCA-based method reacts slower than the DCCA-based one, in addition, the dynamic method gives a higher FDR than the static one. As seen in Figure 7.6, the third fault can be detected effectively by the CCA-based method, but it has smaller FDR compared with the DCCA-based one as shown in Figure 7.7. For the last fault scenario, Figures 7.8 and 7.9 show that both methods can detect the fault. Due to process dynamics, the static method has a large detection delay. Their performances can also be distinguished by comparing the FDR as shown in Table 7.4.

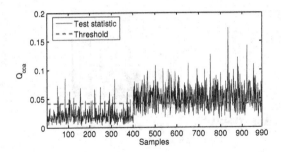

Figure 7.2: Detection result for fault scenario 1 based on CCA

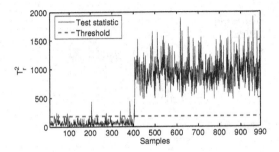

Figure 7.3: Detection result for fault scenario 1 based on DCCA

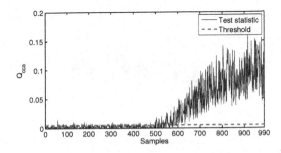

Figure 7.4: Detection result for fault scenario 2 based on CCA

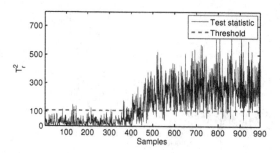

Figure 7.5: Detection result for fault scenario 2 based on DCCA

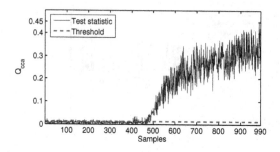

Figure 7.6: Detection result for fault scenario 3 based on CCA

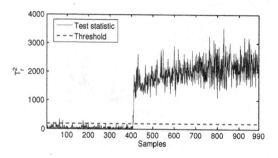

Figure 7.7: Detection result for fault scenario 3 based on DCCA

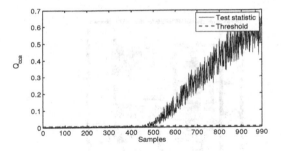

Figure 7.8: Detection result for fault scenario 4 based on CCA

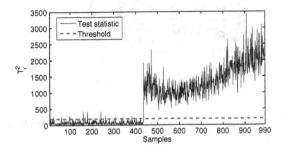

Figure 7.9: Detection result for fault scenario 4 based on DCCA

7.2 Case Study on the CSTH Benchmark Process

7.2.1 Process Description

CSTH is widely used in the chemical industry as a common subsystem. Inside the CSTH, a certain temperature and level of reactants are being held, which define the operating point under which an optimal reaction is possible. The laboratory sized CSTH considered here is a RT 682 CSTH demonstrator plant manufactured by G.U.N.T. Gerätebau GmbH Hamburg[1], as shown in Figure 7.10. The schematic of the CSTH plant is shown in Figure 7.11. Water enters the tank and is heated by a heating jacket. The water then leaves the tank and is recycled to the reservoir. Let $T_i, i = 1, \ldots, 4$ denote the temperature inside the stirred tank, the temperature inside the heating jacket, the input water temperature, and the temperature in the reservoir, respectively; L_1 represent the water level inside the stirred tank; and F_1 denote the cold water flow.

---

[1]The reader is referred to www.gunt.de for details

Figure 7.10: Laboratory setup of CSTH process

7.2.2 Application of Improved Static CCA-based Method

In this study, L_1 is the output variable, T_1 to T_4 and F_1 are the process variables. The sampling period is 1 s. Five hundred samples were collected as the training data set. The control limits were determined at a 0.01 significance level. The moving window length $w_0 = 50$ is determined by off-line tuning. For verification of the proposed approach, two fault scenarios are considered. The description of faults and discussion of the detection results of both methods are given below.

The first fault simulates a change in the heat transfer coefficient to mimic fouling on the surface of the heating coil. Such fouling is common in areas with hard water or dealing with harsh operating conditions involving exotic fluids and extreme temperatures and pressures. In this fault case, the heat transfer coefficient decreases from 100% to 90%. This fault is introduced at the 740th sample and removed at the 1700th sample. Hence, such an event has the potential to go unnoticed by the plant operators. Figure 7.12 shows the detection result. It can be observed that the T_r^2 statistic became sensitive only after about 850 samples. In contrast, the new method can immediately react to this fault and raise an alarm to the plant operators. The second fault occurs in the level sensor, which measures the level inside the stirred tank. In this case, some bubbles entered the plastic hose, which connects the tank and the sensor. As a result, the noise level was increased. The fault was introduced at the 400th sample and removed at the 900th sample. The level measurement for the whole process is given in Figure 7.13, it can be seen that, after the

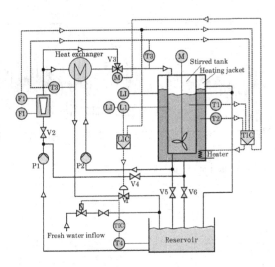

Figure 7.11: The schematic of CSTH process

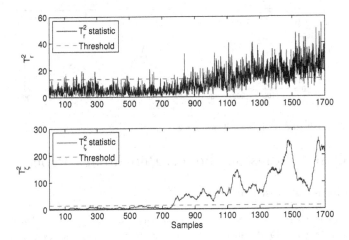

Figure 7.12: Detection results for fault scenario 1 in CSTH

400th sample, the variance tends to be larger. Figure 7.14 show the results for this case. It can be noted that the T_r^2 statistic fails to detect this slight change. However, the proposed method can successfully detect the fault, since the T_ζ^2 statistic of it crosses the threshold.

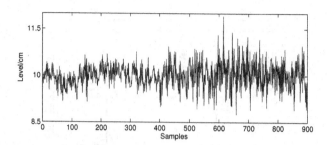

Figure 7.13: Level measurement after fault occurrence

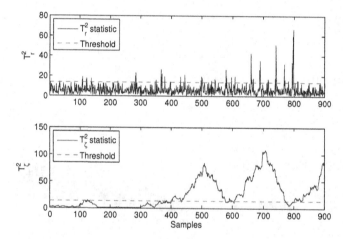

Figure 7.14: Detection results for fault scenario 2 in CSTH

7.3 Case Studies on the TE Benchmark Process

7.3.1 Process Description

The Tennessee Eastman process is designed to simulate a real chemical producing process. As shown in Figure 7.15, the major unit operations are the reactor, the product condenser, the vapour-liquid separator, the recycle compressor and the product stripper. A water-cooling system is used to transfer additional heat in the reactor, since all reactions are exothermic. The three gaseous reactants, A, D and E are fed directly into the reactor. Reactant C and an amount of A (stream 4) enter the process through the stripper. After reacting, the generated products stream leaves the reactor. Subsequently, this stream runs through the condenser where the vaporous components liquify. Then, the product

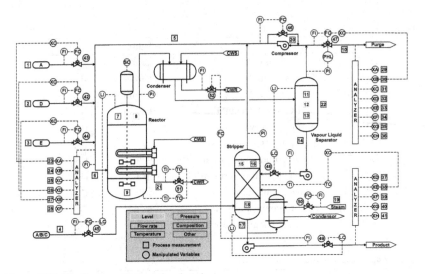

Figure 7.15: The Tennessee Eastman benchmark

stream is further fed to the vapour-liquid separator, the noncondensed components are fed back through the compressor to the reactor feed. In the downstream stripping column, the remaining reactants are removed using the components of feed stream 4 as stripping agents. TE process allows total 52 measurements, including 11 manipulated variables (42)–(52) and 41 process variables ($\boxed{1}$ – $\boxed{41}$) [28]. Technical details of TEP can be found in Ricker [96], where a simulator has also been developed. The simulator is available on-line at the website.[2] Note that there are six operating modes defined by Downs and Fogel [31]. In this study, mode one is used and the decentralized control strategy developed in Ricker [96] is used.

In this work, the manipulated variables are chosen as listed in Table 7.5, while the component F, G and H in purge gas analysis (marker($\boxed{34}$ – $\boxed{36}$)) are applied as the output variables. The proposed method is compared with the standard approach proposed in Chapter 4.

7.3.2 Application of Improved Dynamic CCA-based Method

The data were recorded at a sampling interval of 3 min. A data set consisting of 960 samples in the fault-free case were collected for training. Table 7.6 shows the parameters determined by the original method, where $s_p = s_f$ is predefined, and the process order n is specified using the method in Odiowei and Cao [88]. For the local statistic, the length

---
[2]http://depts.washington.edu/control/LARRY/TE/download.html

Table 7.5: Manipulated variables in TE

Number	Description	Units	Marker
XMV(1)	D feed (stream 2)	kgh^{-1}	㊷
XMV(2)	E feed (stream 3)	kgh^{-1}	㊹
XMV(3)	A feed (stream 1)	kscmh	㊸
XMV(4)	A and C feed (stream 4)	kscmh	㊺
XMV(5)	Compressor recycle valve	%	㊻
XMV(6)	Purge valve (stream 9)	%	㊼
XMV(7)	Separator underflow (stream 10)	m^3h^{-1}	㊽
XMV(8)	Stripper underflow (stream 11)	m^3h^{-1}	㊾
XMV(9)	Stripper steam flow	%	㊿
XMV(10)	Reactor cooling water flow	m^3h^{-1}	(51)
XMV(11)	Condenser cooling water flow	m^3h^{-1}	(52)

Table 7.6: Parameters in the training phase

m	l	s_f	s_p	N	w_0
3	11	5	5	960	60

of moving window $w_0 = 60$ was suggested by off-line tuning. Finally, a significance level of 0.05 was set for the threshold.

One fault-free scenario and three fault scenarios were simulated. Each simulation collects 860 samples, including a set of faulty data from 490 to the end. For performance evaluation, FAR and FDR are used [118]. The higher the FDR, the better is the performance of the corresponding method.

Figure 7.16 shows the fault-free detection results, where it can be observed that T_ζ^2 provides acceptable results with a FAR of less than 5%. This suggests that the proposed method has been properly configured and can follow the normal process operation.

The first fault was simulated by doubling the noise of the reactor's pressure sensor. The achieved results are shown in Figure 7.17, where it can be seen that the T_r^2 statistic performs very poorly, with only 20.8% of the faulty samples detected. In contrast, when using the proposed method, T_ζ^2 can detect the fault well with a FDR of 97.1%.

The second fault simulates a random variation in the D feed temperature. The third fault is simulated as well by a random change in reactor cooling water inlet temperature. Detection results of the two fault scenarios are presented in Figures 7.18 and 7.19. Applying the traditional and proposed approaches to both faults. The detection results show that the conventional DCCA-based method is insensitive to small variations. However, with the proposed approach, the detection performance is greatly improved. It can be noted that the FDR for both methods is close to 99%. Based on the comparison results, it can be seen that the greatest improvement was achieved by introducing the statistical local approach to conventional CCA-based method.

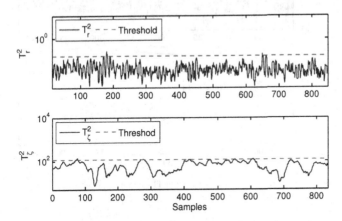

Figure 7.16: Detection results for fault-free scenario in TE process

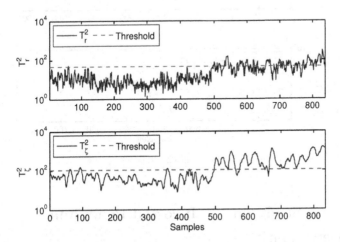

Figure 7.17: Detection results for fault scenario 1 in TE process

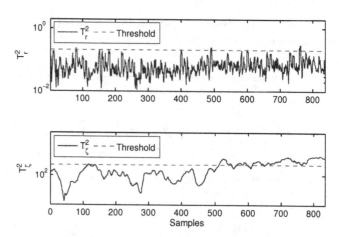

Figure 7.18: Detection results for fault scenario 2 in TE process

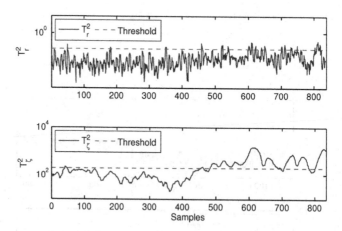

Figure 7.19: Detection results for fault scenario 3 in TE process

7.4 Case Study on the Inverted Pendulum Benchmark

In this section, the application of the proposed method is demonstrated on the inverted pendulum (IP) benchmark.

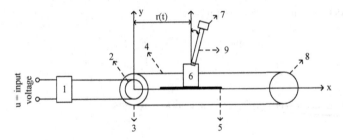

Figure 7.20: Schematic description of an inverted pendulum system

Table 7.7: Measurable variables in IP system

Type of variable	Symbol	Description
Input	u	the acting control voltage
Output	r	the linear position of cart
Output	$\dot{r}$	the velocity of cart
Output	θ	angular position of the pendulum

7.4.1 Benchmark Description

The inverted pendulum system consists of a cart (pos. 6 in Figure 7.20) that moves along a metal guiding bar (pos. 5). An aluminum rod (pos. 9) with a cylindrical weight (pos. 7) is fixed to the cart by an axis. The cart is connected by a transmission belt (pos. 4) to a drive wheel (pos. 3). The wheel is driven by a current controlled direct current motor (pos. 2) that delivers a torque proportional to the acting control voltage u_s such that the cart is accelerated. This system is nonlinear and consists of four state variables:

- the position of the cart r (marked by 6 in Figure 7.20)

- the velocity of the cart $\dot{r}$

- the angle of the pendulum θ as well as

- the angle velocity $\dot{\theta}$

Among the above state variables, r is measured by means of a circular coil potentiometer that is fixed to the driving shaft of the motor, $\dot{r}$ by means of the tacho-generator that is also fixed to the motor and Φ by means of a layer potentiometer fixed to the pivot of the pendulum. The system input u is the acting control voltage u_s that generates force F on the cart. Hence, input and output variables of IP are listed in Table 7.7. There are two types of frictions in the system that affect the system dynamics: dynamic friction and static friction, which are described by

$$\text{dynamic friction: } F_c = -|F_c|sgn(r)$$

$$\text{static friction: } F_{HR} = \begin{cases} -\mu F_n, \dot{r} = 0 \\ 0, \quad \dot{r} \neq 0 \end{cases}$$

To include their effect in the system model, F is extended to

$$F_{sum} = F + d$$

with d being a unknown input. Since this system is nonlinear, the linearization at an operating point can build a linear model.

IP is a one-input and three-output system. Given a sampling time $T = 0.03$ s, we obtain the following linearized discrete-time model of inverted pendulum

$$\mathbf{x}(k+1) = \mathbf{A}_d\mathbf{x}(k) + \mathbf{B}_d\mathbf{u}(k) + \mathbf{E}_d\mathbf{d}(k),$$
$$\mathbf{y}(k) = \mathbf{C}_d\mathbf{x}(k) + \mathbf{v}(k) \tag{7.1}$$

$$\mathbf{A}_d = \begin{bmatrix} 1.0000 & 0.0001 & -0.0569 & 0.0000 \\ 0 & 1.0097 & 0.0116 & 0.0300 \\ 0 & -0.0038 & 0.9442 & -0.0000 \\ 0 & 0.6442 & 0.7688 & 1.0056 \end{bmatrix}$$

$$\mathbf{B}_d = \mathbf{E}_d = \begin{bmatrix} 0.0053 \\ 0.0373 \\ -0.1789 \\ 2.4632 \end{bmatrix}, \mathbf{C}_d = \begin{bmatrix} 1 & 0 & 0 & 0 \\ 0 & 1 & 0 & 0 \\ 0 & 0 & 1 & 0 \end{bmatrix}$$

where $\mathbf{d}(k)$ represents the deterministic disturbance. Because IP is unstable in open-loop, an observer-based state feedback control law is used in this system, which can be found in details in [22].

7.4.2 Simulation Setting

In this simulation, the reference signal is kept constant at $\nu = 0.4$ and the deterministic disturbance is chosen as sinusoidal signal, whose boundedness is unknown. Subsequently, the training data set is collected from previous normal running, which consists of 2000 fault-free samples. The parameters for the proposed method are determined as $s_p=6$, $s_f = 5$ and the length of the evaluation window $N = 500$. Then, the threshold J_{th} is estimated as 0.98×10^{-3}. Three fault scenarios are considered in our study. Since each scenario runs for 39 s, a total of 1300 samples are generated. The considered faults are simulated as

- Fault scenario 1: Offset in the position sensor of the cart. Badly calibrated sensors exist in the practice. We simulate a offset of the position sensor at $t = 30$ s with magnitude equals to 0.15 m

- Fault scenario 2: Offset in the angular position sensor of the pendulum occurs at $t = 30$ s with magnitude equal to 0.20 rad$= 11.46°$

- Fault scenario 3: Offset in the actuator. Actuator faults occur often in practice. The effect of actuator faults can be presented by the variation of the input. In this simulation, the offset actuator fault is introduced at $t = 30$ s with a magnitude equal to 1.6 V

7.4.3 Results and Discussion

Figure 7.21 shows the detection results for fault-free scenario. For comparison, the results of the conventional data-driven PS-based method have been plotted in the top figure. It can be seen that the PS-based method gives large FAR due to the influence of deterministic disturbances. However, the evaluation function J, the bottom one in Figure 7.21, provides acceptable results. This suggests that the proposed method has been properly configured and can follow normal process operation. The detection results for three fault scenarios are shown in Figure 7.22, Figure 7.23 and Figure 7.24, respectively. Each fault is expected to occur at the 1000th sample. Since the threshold represents the maximum influence of the disturbance on measurement, the expected FAR should be zero. From the three figures, we can see that the proposed method has been properly configured and can follow the normal process operation with zero FAR before the onset of fault. Figures 7.22 and 7.23 show the detection result for the first fault and second fault scenarios, respectively. Both faults are sensor faults. The first fault leads to a bias measurement of the position of the cart and the second fault simulates an offset in angular position sensor. It can be seen from the vertical line at the 1000th sample in both figures that the evaluation function rapidly detects this type of fault. The faults are successfully detected only with small detection delays. The detection delay mainly is attributed to the work of controller and the update strategy of evaluation window, in which only one sample leaves and in. It can also be observed that the detection delay in the first figure is less than in the second one due to the fact that the actuator directly works on the cart to change its position, which is measured by the sensor r. The last fault scenario was simulated by an offset in actuator. Figure 7.24 shows the detection result, where the fault is successfully detected.

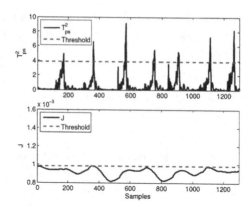

Figure 7.21: Detection results for fault-free scenario

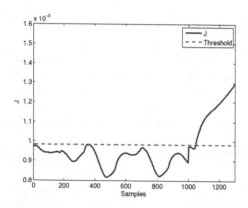

Figure 7.22: Detection result for fault scenario 1

7.5 Concluding Remarks

In this chapter, four benchmarks have been used to demonstrate the performance of the proposed methods. The static and dynamic CCA-based methods in Chapter 4 have been used to detect faults in the alumina evaporation processes. The achieved results have shown that both methods perform well in terms of FAR and MTFA for fault-free scenario and FDR for faulty scenarios, and the dynamic one is superior than the static one in most case studies. In addition, the pilot scale CSTH benchmark has been used to show the effectiveness of the improved static CCA-based FD method. The TE benchmark process, which is dynamic but achieved in steady state, has been utilized to demonstrate

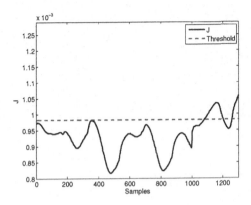

Figure 7.23: Detection result for fault scenario 2

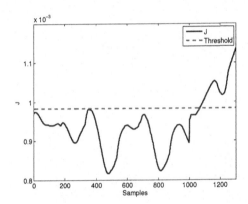

Figure 7.24: Detection result for fault scenario 3

the effectiveness of the improved dynamic CCA-based method. Finally, a simulation on an inverted pendulum system was studied. The achieved results show that the proposed method can address the FD problem with deterministic disturbances well. The case study results illustrate that all tested data-driven methods are applicable in industrial processes.

8 Conclusions and Future Work

8.1 Conclusions

In this dissertation, the evaluation and comparison of two basic detection statistics for data-driven FD methods have been carried out, and advanced data-driven FD methods have been developed for complex industrial processes.

In Chapter 1, the background, basic concepts on fault detection and the motivations of the work have been introduced. On one hand, the intricacies present in modern industrial plants impede application of model- or knowledge-based FD techniques in large-scale processes; on the other hand, techniques for routine data collecting, storing and processing have been significantly improved in past years. Motivated by this, the objectives of this dissertation are first to evaluate and compare two basic detection statistics, and then to develop advanced data-driven FD methods in different application scopes.

Basis of this dissertation has been given in Chapter 2. Following the mathematical description of the static and dynamic processes, the basic principle of FD has been introduced. Subsequently, two commonly used statistics, T^2 and Q, have been discussed and then basic FD approaches have been reviewed. Statistical FD methods including the PCA-, PLS- and DPCA-based methods played essential role in detection of faults in static and steady state dynamic processes. For general dynamic processes, the model-based techniques such as FDF, DO and PS have been introduced and further generalized in the kernel representation framework. Finally, the data-driven kernel representation-based FD methods have been presented.

Although the T^2 and Q statistics have been widely used for data-driven FD methods, there is currently a lack of work aiming at comparing them. To this end, Chapter 3 has evaluated and compared the T^2 statistic and three cases of the Q statistic. For evaluation, a new index, MTFA, has been proposed. Different from FAR, it can tell a process operator how frequently false alarms will occur. Furthermore, the geometric relationship between the T^2 statistic and three cases of the Q statistic, i.e., Q_{max}, Q_{tr} and Q_{gh}, have been discussed. Finally, based on the FD performance with respect to MTFA and FDR given by numerical examples, the recommendation for using the two statistics is that *the T^2 statistic is the better choice for FD purpose, if it is unavailable, the alternative should be the Q_{gh} statistic.*

Based on above comparison, Chapter 4 has focused on advanced FD methods for static and steady state dynamic processes. The most widely used FD methods are PCA- and PLS-based ones. However, the former one often ignores the input-output relationship and the latter one handles the case that output variables are on-line unmeasurable or measurable only with a large time delay. The CCA-based methods can be viewed as an extension of them for detection of faults with available process input and output data. The core of the proposed methods is to find transformations of both process input and output variables and then use achieved transformations to generate residual signals. This can also be regarded as improvements of the conventional methods. Furthermore, compared with the CVA-based methods in dealing with dynamic processes, the new method reduces the engineering effort by avoiding system identification. This characteristic is very convenient for practical applications.

In practice, early detection of multiplicative faults is important. It has shown that CCA-based method is less powerful to deal with incipient multiplicative faults. Consequently, improved CCA-based methods, which incorporate the statistical local approach with the original one, have been developed in Chapter 5. Furthermore, reasons why the T^2 statistic is insensitive to incipient multiplicative faults and why the statistical local approach can make an improvement have been discussed.

In Chapter 6, a novel FD method for dynamic processes has been proposed, in which the so-called residual signals are generated by means of a projection of process input data. This is the major difference to the existing model-based and data-driven FD methods, where residual generator is realized based on the process input and output relationship/dynamics. Furthermore, this way of residual generation circumvents the parameter identification procedure and thus is easier to be used. On the other hand, it also allowed us to address deterministic disturbances, which are often not taken into account by data-driven FD methods. In this fashion, the new method can save the design effort and broaden the application scope.

Finally, the algorithms developed in Chapters 4-6 have been tested on four benchmark cases in Chapter 7. The achieved results have shown that the proposed methods are suitable for practical applications.

8.2 Future Work

Based on the current work, future studies on the following issues would be conducted:

- Nonlinear and non-Gaussian issues: The achieved results in this work are limited to a linear system description or Gaussian distribution. To improve feasibility and effectiveness, more effort has to be made to deal with nonlinear and non-Gaussian issues in large-scale industrial processes.

- Fault isolation and identification issues: To find out the root causes and provide assistance for correct actions, efficient fault isolation and identification techniques would be helpful. Besides the existing work based on contribution plot and reconstruction-based contribution [83, 94, 124], techniques from machine learning and artificial intelligence area [3, 36] are also promising. Combining them with the current work is worth studying. Note that fault isolation is more involved in the closed-loop case.

- Interconnection between CCA and model-based methods: Although both types of methods stem from different motivations, studying their possible interconnection would be helpful to the integration of data-driven and model-based methods [24].

Bibliography

[1] Data, data everywhere. *The Economist*, 2010.

[2] H. Akaike. Stochastic theory of minimal realization. *IEEE Transactions on Automatic Control*, 19(6):667–674, 1974.

[3] C. Aldrich and L. Auret. *Unsupervised Process Monitoring and Fault Diagnosis with Machine Learning Methods*. Springer-Verlag, London, 2013.

[4] S. T. Alexander. *Adaptive Signal Processing: Theory and Applications*. Springer-Verlag, New York, 1986.

[5] T.W. Anderson. *An Introduction to Multivariate Statistical Analysis*. John Wiley and Sons, LTD, New York, second edition, 1984.

[6] K. J. Aström. *Introduction to Stochastic Control Theory*. Academic Press, 1970.

[7] M. Basseville. Information criteria for residual generation and fault detection and isolation. *Automatica*, 33(5):783 – 803, 1997.

[8] M. Basseville. On-board component fault detection and isolation using the statistical local approach. *Automatica*, 34(11):1391–1415, 1998.

[9] M. Basseville and I.V. Nikiforov. *Detection of Abrupt Changes: Theory and Application*. Prentice-Hall, New York, 1993.

[10] B. Boulkroune, M. Galvez-Carrillo, and M. Kinnaert. Additive and multiplicative fault diagnosis for a doubly-fed induction generator. In *Proceedings of the 2011 IEEE International Conference on Control Applications (CCA)*, pages 1302–1308, Denver, USA, 2011.

[11] G. E. P. Box. Some theorems on quadratic forms applied in the study of analysis of variance problems: Effect of inequality of variance in one-way classification. *Annals of Mathematical Statistics*, 25:290–302, 1954.

[12] G. E. P. Box, S. Graves, S. Bisgaard, J. Van Gilder, K. Marko, J. James, M. Seifer, M. Poublon, and F. Fodale. Detecting malfunctions in dynamic systems. Technical report, SAE, 2000.

[13] CCPS (Center for Chemical Process Safety). *Recognizing Catastrophic Incident Warning Signs in the Process Industries*. John Wiley and Sons, Inc., 2012.

[14] Q. Q. Chai, C. H. Yang, K. L. Teo, and W. H. Gui. Optimal control of an industrial-scale evaporation process: Sodium aluminate solution. *Control Engineering Practice*, 5(5):663–670, 2012.

[15] Z. W. Chen, S. X. Ding, K. Zhang, Z. B. Li, and Z. K. Hu. Canonical correlation analysis-based fault detection methods with application to alumina evaporation process. *Control Engineering Practice*, 46:51–58, 2016.

[16] Z. W. Chen, K. Zhang, S. X. Ding, , Y. A. W. Shardt, and Z. K. Hu. Improved canonical correlation analysis-based fault detection methods for industrial processes. *Jouranl of Process Control*, 41:26–34, 2016.

[17] Z. W. Chen, K. Zhang, S. X. Ding, X. Yang, Z. M. He, and Z. K. Hu. Study on small multiplicative fault detection using canonical correlation analysis with the local approach. In *Proceedings of the 9th IFAC Symposium on SAFEPROCESS*, pages 1414–1419, Paris, France, 2015.

[18] L. H. Chiang, E. L. Russell, and R. D. Braatz. Fault diagnosis in chemical processes using fisher discriminant analysis, discriminant partial least squares, and principal component analysis. *Chemometrices and Intelligent Laboratory Systems*, 50:243–252, 2000.

[19] L. H. Chiang, E. L. Russell, and R. D. Braatz. *Fault Detection and Diagnosis in Industrial Systems*. Springer-Verlag, London, 2001.

[20] S. W. Choi, C. Lee, J. M. Lee, J. Park, and I. B. Lee. Fault detection and identification of nonlinear processes based on kernel PCA. *Chemometrics and Intelligent Laboratory Systems*, 75(1):55–67, 2005.

[21] E. Chow and A.S. Willsky. Analytical redundancy and the design of robust failure detection systems. *IEEE Transactions on Automatic Control*, 29(7):603–614, 1984.

[22] S. X. Ding. *Model-Based Fault Diagnosis Techniques—Design Schemes, Algorithms and Tools, 2nd ed*. Springer-Verlag, London, 2013.

[23] S. X. Ding. *Data-driven Design of Fault Diagnosis and Fault-tolerant Control Systems*. Springer-Verlag, London, 2014.

[24] S. X. Ding. Data-driven design of monitoring and diagnosis systems for dynamic process, a review of subspace technique based schemes and some recent results. *Journal of Process Control*, 24:431–449, 2014.

[25] S. X Ding. Application of factorization and gap metric techniques to fault detection and isolation Part I: A factorization technique based FDI framework. In *Proceedings of the 9th IFAC Symposium on SAFEPROCESS*, pages 119–124, Paris, France, September 2015.

[26] S. X. Ding, S. Yin, K. X. Peng, H. Y. Hao, and B. Shen. A novel scheme for key performance indicator prediction and diagnosis with application to an industrial hot strip mill. *IEEE Transactions on Industrial Informatics*, 9(4):2239–2247, 2013.

[27] S. X. Ding, P. Zhang, E. L. Ding, S. Yin, A. Naik, P. C. Deng, and W. H. Gui. On the application of PCA technique to fault diagnosis. *Tsinghua Science and Technology*, 15:138–144, 2010.

[28] S. X. Ding, P. Zhang, A. Naik, E. L. Ding, and B. Huang. Subspace method aided data-driven design of fault detection and isolation systems. *Journal of Process Control*, 19:1496–1510, 2009.

[29] J. F. Dong and M. Verhaegen. Subspace based fault detection and identification for LTI systems. In *Proceedings of the 7th IFAC Symposium on SAFEPROCESS*, pages 330–335, Barcelona, Spain, 2009.

[30] J. F. Dong, M. Verhaegen, and F. Gustafsson. Robust fault detection with statistical uncertainty in identified parameters. *IEEE Transactions on signal processing*, 60:5064–5076, 2012.

[31] J.J. Downs and E.F. Fogel. A plant-wide industrial process control problem. *Computers & Chemical Engineering*, 17:245–255, 1993.

[32] L. M. Elshenawy, S. Yin, A. S. Naik, and S. X. Ding. Efficient recursive principal component analysis algorithms for process monitoring. *Industrial & Engineering Chemistry Research*, 49:252–259, 2010.

[33] D. Fenna. *Cartographic Science: A Compendium of Map Projections, with Derivations*. CRC, 2006.

[34] P. M. Frank. Fault diagnosis in dynamic systems using analytical and knowledge-based redundancy-a survey and some new results. *Automatica*, 26:459–474, 1990.

[35] Z. Q. Ge and Z. H. Song. Process monitoring based on independent component analysis-principal component analysis (ICA-PCA) and similarity factors. *Industrial & Engineering Chemistry Research*, 46:2054–2063, 2007.

[36] Z. Q. Ge, Z. H. Song, and F. R. Gao. Review of recent research on data-based process monitoring. *Industrial & Engineering Chemistry Research*, 52(10):3543–3562, 2013.

[37] Z. Q. Ge, L. Xie, U. Kruger, and Z. H. Song. Local ICA for multivariate statistical fault diagnosis in systems with unknown signal and error distributions. *AIChE Journal*, 58(8):2357–2372, 2012.

[38] J. J. Gertler. *Fault Detection and Diagnosis in Engineering Systems*. Marcel Dekker, New York., 1998.

[39] W. Gilchrist. *Statistical Modelling with Quantile Functions*. CRC Press, 2000.

[40] F. Gustafsson. *Adaptive Filtering and Change Detection*. John Wiley and Sons, LTD, 2000.

[41] A. Hagahni. *Data-Driven Design of Fault Diagnosis Systems for Nonlinear Multimode Processes*. PhD thesis, University of Duisburg-Essen, Germany, 2013.

[42] Ch. Hajiyev. Tracy–widom distribution based fault detection approach: Application to aircraft sensor/actuator fault detection. *ISA Transactions*, 51(1):189–197, 2012.

[43] H. Y. Hao. *Key Performance Monitoring and Diagnosis in Industrial Automation Processes*. PhD thesis, University of Duisburg-Essen, 2014.

[44] H. Y. Hao, S. X. Ding, A. Haghani, S. Yin, and T. Jeinsch. Influence of additive and multiplicative faults on process output variances. In *Proceedings of the 1st PAPYRUS Workshop on Fault Diagnosis and Fault Tolerant Control in Large Scale Processing Monitoring*, 2011.

[45] H. Y. Hao, K. Zhang, S. X. Ding, Z. W. Chen, and Y. G. Lei. A data-driven multiplicative fault diagnosis approach for automation processes. *ISA Transactions*, 53(5):1436 – 1445, 2014.

[46] W. K. Härdle and L. Simar. *Applied Multivariate Statistical Analysis, Third Edition*. Springer-Verlag, Berlin Heidelberg, 2012.

[47] J. Harmouche, C. Delpha, and D. Diallo. Incipient fault detection and diagnosis based on kullback-leibler divergence using principal component analysis: Part I. *Signal Processing*, 94:278 – 287, 2014.

[48] D. M. Hawkins. Multivariate quality control based on regression-adjusted variables. *Technometrics*, 33(1):pp. 61–75, 1991.

[49] H. Hotelling. Relations between two sets of variates. *Biometrika*, 28(3-4):321–377, 1936.

[50] Z. K. Hu, Z. W. Chen, W. H. Gui, and B. Jiang. Adaptive PCA based fault diagnosis scheme in imperial smelting process. *ISA Transactions*, 53(5):1446–1455, 2014.

[51] Z. K. Hu, Z. W. Chen, C. C. Hua, W. H. Gui, and S. X. Ding. A simplified recursive dynamic pca based monitoring scheme for imperal smelting process. *International Journal of Innovative Computing, Information and Control*, 8(4):2551–2561, 2012.

[52] B. Huang and R. Kadali. *Dynamic Modelling, Predictive Control and Performance Monitoring, a Data-Driven Subspace Approach*. Springer-Verlag, London, 2008.

[53] R. Isermann. *Fault Diagnosis Systems*. Springer-Verlag, London, 2006.

[54] J. E. Jackson. Quality control methods for several related variables. *Technometrics*, 1:359 – 377, 1959.

[55] J. E. Jackson and R. H. Morris. An application of multivaraite quality control to photographic processing. *Journal of the American Statistical Association*, 52(278):186–199, 1957.

[56] J. E. Jackson and G. S. Mudholkar. Control procedures for residuals associated with principal component analysis. *Technometrics*, 21:341–349, 1979.

[57] P.A. Janakiraman and S. Renganathan. Recursive computation of pseudo-inverse of matrices. *Automatica*, 18(5):631 – 633, 1982.

[58] I.T. Jolliffe. *Principal Component Analysis*. Springer-Verlag, New York, Berlin, 1986.

[59] T. Kailath, A.H. Sayed, and B. Hassibi. *Linear Estimation*. Prentice Hall, New Jersey, 1999.

[60] P. K. Kankar, S. C. Sharma, and S. P. Harsha. Rolling element bearing fault diagnosis using wavelet transform. *Neurocomputing*, 74(10):1638 – 1645, 2011.

[61] M. Kano, S. Hasebe, I. Hashimoto, and H. Ohno. A new multivariate statistical process monitoring method using principal component analysis. *Computers & Chemical Engineering*, 25(7-8):1103 – 1113, 2001.

[62] T. Katayama and G. Picci. Realization of stochastic systems with exogenous inputs and subspace identification methods. *Automatica*, 35(10):1635–1652, 1999.

[63] K. Kim, J. M. Lee, and I. B. Lee. A novel multivariate regression approach based on kernel partial least squares with orthogonal signal correction. *Chemometrics and Intelligent Laboratory Systems*, 79(1-2):22–30, 2005.

[64] T. Komulainen, M. Sourander, and S. L. Jamsa-Jounela. An online application of dynamic PLS to a dearomatization process. *Computers & Chemical Engineering*, 28(12):2611–2619, 2004.

[65] J. V. Kresta, J. F. MacGregor, and T. E. Marlin. Multivariate statistical monitoring of process operating performance. *The Canadian Journal of Chemical Engineering*, 69(1):35–47, 1991.

[66] U. Kruger, S. Kumar, and T. Littler. Improved principal component monitoring using the local approach. *Automatica*, 43(9):1532 – 1542, 2007.

[67] U. Kruger and L. Xie. *Advances in Statistical Monitoring of Complex Multivariate Processes: With Applications in Industrial Process Control.* Wiley, 2012.

[68] W. F. Ku, R. H. Storer, and C. Georgakis. Disturbance detection and isolation by dynamic principal component analysis. *Chemometrics and Intelligent Laboratory Systems*, 30(1):179–196, 1995.

[69] W. E. Larimore. System identification reduced order filtering and modeling via canonical variate analysis. In *Proceedings of the American Control Conference*, pages 175–181, San Francisco, USA, 1983.

[70] J. M. Lee, S. J. Qin, and I. B. Leen. Fault detection and diagnosis based on modified independent component analysis. *AIChE Journal*, 52(10):3501–3514, 2006.

[71] Y. G. Lei, J. Lin, Z. J. He, and M. J. Zuo. A review on empirical mode decomposition in fault diagnosis of rotating machinery. *Mechanical Systems and Signal Processing*, 35(1-2):108 – 126, 2013.

[72] Y. G. Lei, J. Lin, M. J. Zuo, and Z. J. He. Condition monitoring and fault diagnosis of planetary gearboxes: A review. *Measurement*, 48:292 – 305, 2014.

[73] W. Li, Z. G. Han, and S. L. Shah. Subspace identification for FDI in systems with non-uniformly sampled multirate data. *Automatica*, 42:619–627, 2006.

[74] S. Liisa and J. Jounela. Future trends in process automation. *Annual Reviews in Control*, 31:211–220, 2007.

[75] H. Luo, S. X. Ding, K. Zhang, and S. Yin. A data-driven fault detection approach for static processes with deterministic disturbances. In *Proceedings of the 23rd IEEE International Symposium on Industrial Electronics (ISIE)*, pages 2404–2409, Istanbul, Turkey, 2014.

[76] J. F. MacGregor, C. Jaeckle, C. Kiparissides, and M. Koutoudi. Process monitoring and diagnosis by multiblock PLS methods. *AIChE Journal*, 40(5):826–838, 1994.

[77] J. F. MacGregor and T. Kourti. Statistical process control of multivariate processes. *Control Engineering Practice*, 3:403–414, 1995.

[78] K. V. Mardia, J. T. Kent, and J. M. Bibby. *Multivariate Analysis*. Academic Press, 1979.

[79] K. F. Martin. A review by discussion of condition monitoring and fault diagnosis in machine tools. *International Journal of Machine Tools and Manufacture*, 34(4):527 – 551, 1994.

[80] M. R. Maurya, R. Rengaswamy, and V. Venkatasubramanian. Application of signed digraphs-based analysis for fault diagnosis of chemical process flowsheets. *Engineering Applications of Artificial Intelligence*, 17(5):501–518, 2004.

[81] P. D. Mcfadden and M. M. Toozhy. Application of synchronous averaging to vibration monitoring of rolling element bearings. *Mechanical Systems and Signal Processing*, 14(6):891–906, 2000.

[82] J. McNames. Fourier series analysis of epicyclic gearbox vibration. *Journal of Vibaration and Acoustics*, 124:150–152, 2001.

[83] P. Miller, R. E. Swanson, and C. E. Heckler. Contribution plots: a missing link in multivariate quality control. *Applied Mathematics and Computer Science*, 8:775–792, 1998.

[84] Robb J. Muirhead. *Aspects of Multivariate Statistical Theory*. Wiley, New York, 1982.

[85] A. Naik. *Subspace Based Data-driven Designs of Fault Detection Systesms*. PhD thesis, University of Duisburg-Essen, Germany, 2010.

[86] NASA. *Fault Tree Handbook with Aerospace Applications*. NASA Office of Safety and Mission Assurance, Washington D.C., 2002.

[87] A. Negiz and A. Cinar. Statistical monitoring of multivariable dynamic process with state space models. *AIChE Journal*, 8:2002–2020, 1997.

[88] P. Odiowei and Y. Cao. Nonlinear dynamic process monitoring using canonical variate analysis and kernel density estimations. *IEEE Transactions on Industrial Informatics*, 6:36–45, 2010.

[89] R. J. Patton and J. Chen. Optimal unknown input distribution matrix selection in robust fault diagnosis. *Automatica*, 29:837–841, 1993.

[90] R. J. Patton, P. M. Frank, and R. N. Clark. *Issues of Fault Diagnosis for Dynamic Systems*. Springer, 2000.

[91] K. X. Peng, K. Zhang, G. Li, and D. H. Zhou. Contribution rate plot for nonlinear quality-related fault diagnosis with application to the hot strip mill process. *Control Engineering Practice*, 21:360–369, 2013.

[92] H. Vincent Poor and Olympia Hadjiliadis. *Quickest Detection.* Cambridge University Press, 2008. Cambridge Books Online.

[93] S. J. Qin. An overview of subspace identification. *Computers and Chemical Engineering*, 30:1502–1513, 2006.

[94] S. J. Qin. Survey on data-driven industrial process monitoring and diagnosis. *Annual Reviews in Control*, 36(2):220–234, 2012.

[95] M. M. Rashid and J. Yu. A new dissimilarity method integrating multidimensional mutual information and independent component analysis for non-gaussian dynamic process monitoring. *Chemometrices and Intelligent Laboratory Systems*, 115:44–58, 2012.

[96] N. Ricker. Decentralized control of the tennessee eastmann challenge process. *Journal of Process Control*, 6:205–221, 1996.

[97] E. L. Russell, L. H. Chiang, and R. D. Braatz. *Data-driven Techniques for Fault Detection and Diagnosis in Chemical Processes.* Springer-Verlag, London, 2000.

[98] SAE. Sae air4985: A methodology for quantifying the performance of an engine monitoring system, 2012.

[99] Y. A. W. Shardt. *Statistics for Chemical and Process Engineers: A Modern Approach.* Springer, Cham, Switzerland, 2015.

[100] S. Simani, S. Fantuzzi, and R. J. Patton. *Model-Based Fault Diagnosis in Dynamic Systems Using Identification Techniques.* Springer-Verlag, 2003.

[101] A. Simoglou, E.B. Martin, and A.J. Morris. Statistical performance monitoring of dynamic multivariate processes using state space modelling. *Computers and Chemical Engineering*, 26:909–920, 2002.

[102] R. Tempo, G. Calafiro, and F. Dabbene. *Randomized Algorithms for Analysis and Control of Uncertain Systems.* Springer, 2005.

[103] N. F. Thornhill and A. Horch. Advances and new directions in plant-wide disturbance detection and diagnosis. *Control Engineering Practice*, 15(10):1196 – 1206, 2007.

[104] N.D. Tracy, J.C. Yong, and R.L. Mason. Multivariate control charts for individual observations. *Journal of Quality Control*, 24:88–95, 1992.

[105] H. Vedam and V. Venkatasubramanian. Signed digraph based multiple fault diagnosis. *Computers & Chemical Engineering*, 21, Supplement:S655–S660, 1997.

[106] V. Venkatasubramanian, R. Rengaswamy, S.N. Kavuri, and K. Yin. A review of process fault detection and diagnosis Part II: Qualitative models and search strategies. *Computers and Chemical Engineering*, 27(3):313–326, 2003.

[107] V. Venkatasubramanian, R. Rengaswamy, S.N. Kavuri, and K. Yin. A review of process fault detection and diagnosis Part III: Process history based methods. *Computers and Chemical Engineering*, 27:327–346, 2003.

[108] V. Venkatasubramanian, R. Rengaswamy, K. Yin, and S.N. Kavuri. A review of process fault detection and diagnosis Part I: Quantitative model-based methods. *Computers & Chemical Engineering*, 27:293–311, 2003.

[109] J. Wang and S. J. Qin. A new subspace identification approach based on principle component analysis. *Journal of process control*, 12:841–855, 2002.

[110] X. Wang, U. Kruger, and B. Lennox. Recursive partial least squares algorithms for monitoring complex industrial processes. *Control Engineering Practice*, 11:613–632, 2003.

[111] Y. Wang, D. E. Seborg, and W. E. Larimore. Process monitoring using canonical variate analysis and principal component analysis. In *IFAC ADCHEM'97*, Banff, Alberta, Canada, June 1997.

[112] Y. L. Wang, H. F. Qiu, W. H. Gui, and C. H. Yang. Alumina evaporation process simulation system based on sub-flowsheet. *Computer Engineering*, 35(23):260–262, 2009.

[113] B. M. Wise and N. B. Gallagher. The process chemometrics approach to process monitoring and fault detection. *Journal of Process Control*, 6(6):329–348, 1996.

[114] H. Wold. *Soft Modeling by Latent Variables; the Nonlinear Iterative Partial Least Squares Approach*. Academic Press, London, 1975.

[115] S. Yang and Q. Zhao. Probability distribution characterisation of fault detection delays and false alarms. *IET Control Theory Applications*, 6(7):953–962, 2012.

[116] Y. Yao, T. Chen, and F. R. Gao. Multivariate statistical monitoring of two-dimensional dynamic batch processes utilizing non-Gaussian information. *Journal of Process Control*, 20(10):1188–1197, 2010.

[117] S. Yin. *Data-Driven Design of Fault Diagnosis Systems*. PhD thesis, University of Duisburg-Essen, 2012.

[118] S. Yin, S. X. Ding, A. Haghani, H. Y. Hao, and P. Zhang. A comparison study of basic data-driven fault diagnosis and process monitoring methods on the benchmark tennessee eastman process. *Journal of Process Control*, 22:1567–1581, 2012.

[119] S. Yin, S. X. Ding, X. C. Xie, and H. Luo. A review on basic data-driven approaches for industrial process monitoring. *IEEE Transactions on Industrial Electronics*, 61(11):6418–6428, 2014.

[120] J. Yu. A support vector clustering-based probabilistic method for unsupervised fault detection and classification of complex chemical processes using unlabeled data. *AIChE Journal*, 59:407–419, 2013.

[121] J. Zeng, U. Kruger, J. Geluk, X. Wang, and L. Xie. Detecting abnormal situations using the kullback-leibler divergence. *Automatica*, 50(11):2777–2786, 2014.

[122] K. Zhang, H. Y. Hao, Z. W. Chen, S. X. Ding, and K. X. Peng. A comparison and evaluation of key performance indicator-based multivariate statistics process monitoring approaches. *Journal of Process Control*, 33:112–126, 2015.

[123] K. Zhang, Y. W. Shardt, Z. W. Chen, S. X. Ding, and K. X. Peng. Unit-level modelling for KPI of batch hot strip mill process using dynamic partial least squares. In *Proceddings of the 17th IFAC Symposium on System Identification*, pages 19–21, Beijing, China, 2015.

[124] P. Zhang, T. Jeinsch, and S. X. Ding. Process monitoring and fault diagnosis-status and applications. In *Proceedings of the 18th IFAC World Congress*, pages 12401–12406, Milano, Italy, 2011.

[125] P. Zhang, H. Ye, S. X. Ding, G. Z. Wang, and D. H. Zhou. On the relationship between parity space and H2 approaches to fault detection. *Systems and Control Letters*, 2006.

[126] Q. Zhang, M. Basseville, and A. Benveniste. Early waring of slight changes in systems. *Automatica*, 30:95–114, 1994.

Printed in the United States
By Bookmasters